Biodiversity of Fishes in Arunachal Himalaya

Systematics, Classification, and Taxonomic Identification

Biodiversity of Fishes in Arunachal Himalaya

Systematics, Classification, and Taxonomic Identification

Achom Darshan
Postdoctoral Research Associate,
Center with Potential for Excellence in Biodiversity: Phase-II,
Department of Zoology, Rajiv Gandhi University, Itanagar, India

Santoshkumar Abujam
Postdoctoral Research Associate,
Department of Zoology, Rajiv Gandhi University, Itanagar, India

Debangshu Narayan Das
Professor, Department of Zoology,
Rajiv Gandhi University, Itanagar, India

ELSEVIER

ACADEMIC PRESS
An imprint of Elsevier

Academic Press is an imprint of Elsevier
125 London Wall, London EC2Y 5AS, United Kingdom
525 B Street, Suite 1650, San Diego, CA 92101, United States
50 Hampshire Street, 5th Floor, Cambridge, MA 02139, United States
The Boulevard, Langford Lane, Kidlington, Oxford OX5 1GB, United Kingdom

Notices
Knowledge and best practice in this field are constantly changing. As new research and experience broaden our understanding, changes in research methods, professional practices, or medical treatment may become necessary.

Practitioners and researchers must always rely on their own experience and knowledge in evaluating and using any information, methods, compounds, or experiments described herein. In using such information or methods they should be mindful of their own safety and the safety of others, including parties for whom they have a professional responsibility.

To the fullest extent of the law, neither the Publisher nor the authors, contributors, or editors, assume any liability for any injury and/or damage to persons or property as a matter of products liability, negligence or otherwise, or from any use or operation of any methods, products, instructions, or ideas contained in the material herein.

Library of Congress Cataloging-in-Publication Data
A catalog record for this book is available from the Library of Congress

British Library Cataloguing-in-Publication Data
A catalogue record for this book is available from the British Library

ISBN 978-0-12-815556-1

Publisher: Andre Gerhard Wolff
Acquisition Editor: Anna Valutkevich
Editorial Project Manager: Jaclyn A Truesdell
Production Project Manager: Sreejith Viswanathan
Cover Designer: Miles Hitchens

Typeset by SPi Global, India

Contents

About the Authors

Dr. Achom Darshan has been working in the field of freshwater fish taxonomy for the last 14 years after the completion of his Master's degree in Zoology with special paper on fish and fisheries. He obtained a degree in Doctor of Philosophy in science from Manipur University, Canchipur for his thesis titled "Phylogeny of the Siluroid fishes of Manipur" under the direct supervision of Prof. W. Vishwanath. Dr. Darshan has currently been pursuing postdoctoral research as a research associate in "Center with Potential for Excellence in Biodiversity" (sponsored by UGC, New Delhi) from Rajiv Gandhi University, Itanagar since 2013. Previously, he was also engaged as Postdoctoral Research Associateship sponsored by DBT, New Delhi under the ICAR-Directorate of Coldwater Fisheries Research, Bhimtal in 2009 and his assignment was continued till 2012 with Intellectual Technology Management Unit under the same directorate. By virtue of his deep involvement and long research experience, Dr. Darshan has been able to discover more than 10 new freshwater fishes from Northeast India and also published more than 25 research articles in various reputed national and international journals.

Dr. SantoshKumar Abujam obtained Master's degree in Life Sciences (Zoology) with special paper on fish and fisheries in 2006 from Dibrugarh University, Assam. He was awarded degree in Doctor of Philosophy in science by the Dibrugarh University in 2011 for his research titled "Studies on the ecology, biology and rearing feasibility of spiny eel *Macrognathus aral* and *Macrognathus pancalus* from upper Assam" under direct supervision of Prof. S.P. Biswas. Dr. Abujam has acquired long experience in diverse areas of fisheries research through several predoctoral and postdoctoral assignments, respectively, as a Project Fellow in a project of ICAR-NBFGR Lucknow; Junior Research Fellow in a project of MoEF, New Delhi; research associate in a project of DBT, New Delhi under the Department of Life Sciences, Dibrugarh University, Assam. Currently, he has been pursuing Postdoctoral Research in a DBT sponsored project as a research associate under the Department of Zoology, Rajiv Gandhi University, Itanagar. By virtue of his deep involvement in the subject, Dr. Abujam was able to publish more than 55 research articles in various national and international journals and 7 articles in book chapter/proceedings.

Prof. Debangshu Narayan Das in the Department of Zoology, Rajiv Gandhi University (formerly Arunachal University) has made significant contribution in the field of mountain fisheries, especially in the area of taxonomy, reproductive biology, and the artificial propagation of hill stream food and ornamental fishes of Arunachal Pradesh since joining here in 1998. Prof. Das has well-organized fisheries research unit and is involved in discovery of new species, delineation of aquatic ecology of hill stream as well as development of culture-based fisheries in the state of Arunachal Pradesh. So far, 10 new species with new nomenclature added to the world record have been discovered by the unit. In addition, he has successfully supervised the works of 12 PhD awardees from the University and currently 2 postdoctoral research associates (RA) and 8 research scholars have been working for research and development in various aspects of eastern Himalayan mountain fisheries. Prof. Das has also successfully completed various research projects under the funding of DBT New Delhi, UGC New Delhi and collaborative programs with ICAR, New Delhi along with an institutional mega projects (sponsored by UGC New Delhi) on CPEB I & II. He published more than 115 research articles in reputed international and national journals, 4 coauthored books, numbers of book chapters, and more than 35 popular articles in different languages.

Foreword

A number of excellent Indian books on "**fish taxonomy**" as well as "**Ichthyology**" are currently available, but to date none have been solely dedicated to the **Arunachal Himalaya**. The works done by the authors are convincingly remarkable as it has covered a strategic land area of the Indian zone of the Himalaya with variations in topography and altitude forming diverse ecological habitats. Arunachal Pradesh, the easternmost Himalayan province of India being located within latitude 26°28′ and 29°30′N, longitude 91°20′ and 97°30′E having a huge forest cover and a profuse rate of annual precipitations (2818.02 mm) form a major part of eastern part of the Himalayan biodiversity hot spot with numbers of endemic fauna and flora. Further, the varied freshwater habitats constituted by rivers, streams, lakes, reservoirs, natural ponds in different altitudes have cradled the pertinent dwelling spaces for the numerous native ichthyofauna. The authors of the book titled "**Biodiversity of Fishes in Arunachal Himalaya**" have described all the living species of fishes meticulously on the backdrop of state's unique ecological settings from different waterbodies as the outcome of their systematic ichthyofaunal explorations. A reading of this valuable document shows it to be quite complete and extremely innovative in its concept for fishery personnel, biodiversity specialists, fish conservationists, and policy makers. Straightforwardly, due to the inclusion of recent discoveries of several new species of fishes, it may be a guiding document, the only one of its kind for researchers that undoubtedly will cater the conservation attempts of precious genetic resources for our posterity. So, it is my proud privilege to thank all the authors of the book and I personally think that this information-rich book will be very valuable reference document for freshwater fish researchers of the country as a whole.

Tamo Mibang

Vice-Chancellor, Rajiv Gandhi University, Doimukh, India

Arunachal Pradesh, an integral part of the Eastern Himalaya region has highly diverse freshwater fish diversity which is of great regional importance to livelihoods and socio-economic status of the people. Some of the diminutive colourful hill stream fishes have great potentials to be adopted as ornamental fish of high demand in the international market. Development activities do not often consider conservation of biodiversity. Thus, adequate representation of the biodiversity and awareness for the protection of the ecosystem for conservation of the endemic and threatened taxa are very much required.

Nath and Dey's (2000) *Fish and Fisheries of Northeast India (Arunachal Pradesh)* was a useful compilation which included brief descriptions of 131 fish species. Research on the fish diversity and systematics with classical and modern tools continues and has brought changes in the systematic positions and nomenclature of several taxa. It has resulted in great confusions to the students and researchers with the old literature. Survey and discovery of fish species of the region gained momentum in the last few decades. A. Darshan, A. Santosh, and D.N. Das have significantly contributed to the field recently. The discovery of a rare glyptosternine, *Creteuchiloglanis payjab*, which was published in *Aqua* and a Siluroid, *Amblyceps waikhomi* in *PLOS One* are noteworthy.

The present book includes diagnostic characters, local names, colour descriptions and key to identification of 218 fish species belonging to 99 genera, 35 families and 13 orders. Up-to-date methodology is adopted; collection data of the examined specimens are provided for further examination by other workers; excellent colour photographs of the species are presented. The compilation is indeed an excellent contribution to Indian Ichthyology which will be very useful to fish taxonomists of the world.

W. Vishwanath

Dean, School of Life Sciences, Manipur University, Canchipur, India

IUCN-South Asia Regional Chair, Freshwater Fish Specialist Group

(South Asia office: Zoo Outreach Organization, Coimbatore: 641035, India)

Preface

This book "**Biodiversity of Fishes in Arunachal Himalaya**" is a modest attempt to document the ichthyofaunal diversity of Arunachal Pradesh, India. We have a great pleasure in presenting this book documenting a total of 218 fish species in the state. These species belong to 99 genera, 35 families covering 13 orders of freshwater fishes. The book has been prepared by incorporating color photographs of live or freshly preserved specimens and images of "types" and "topotypes" of recently described species. This book provides the summary of order, generic characters, key to species, key diagnostic features, vernacular names, conservational status, and coloration of fresh or preserved specimens along with their distribution patterns in lotic and lentic waterbodies. The present document has been updated on the backdrop of specimens preserved and registered mainly in the Rajiv Gandhi University Museum of Fishes (RGUMF), Doimukh; Arunachal Pradesh Regional Centre, Zoological Survey of India (APRC-ZSI), Itanagar; Manipur University Museum of Fishes, Canchipur, Manipur; and Lakhimpur Girls' College Fish Museum, Lakhimpur, Assam.

During the past two decades, ichthyologists showed keen interest in the exploration of piscatorial resources and accordingly many new fish taxa have been discovered from the state till date. In fact, documentation process is continuing because many remote and poorly communicated areas are yet to be covered. Further, certain endemic fishes are gradually being exposed to the increasing threat of habitat shrinkages and anthropocentric developmental activities.

The taxonomic reassessment and classification used in this book has been followed in accordance with the recently amended taxonomic concept of the International Code of Zoological Nomenclature. The book covers all the food and ornamental fishes dwelling in topographical zones comprising cold waters in high altitudes, cold and warm waters in mid-altitude, and warm waters in low-altitude foothills. We believe that this book will provide an in-depth understanding of ichthyofaunal resources of this mountainous state of northeast India and may help in accurate assessments of fish wealth for enhancing conservation, utilization, management, and ultimately the productivity from such resources by the people of the state. Moreover, this work may pave the way to a new vista toward strengthening hill aquaculture through domestication of wild food and ornamental fishes for the welfare of mountain inhabitants.

Further, this book will serve as a comprehensive field guide for instant identification of freshwater fishes for the students, researchers, amateur naturalists, ichthyologists, and fisher folks in general.

Achom Darshan
Santoshkumar Abujam
Debangshu Narayan Das

Acknowledgments

The authors are grateful to Prof. Tamo Mibang, Vice Chancellor of Rajiv Gandhi University, Rono Hills, Doimukh for his continues support in fisheries research especially in the mountain fish diversity of Arunachal Pradesh, India.

The authors are indebted to the heads of the Department of Zoology, RGU for their unstinted help from time to time in allowing all the available infrastructural facilities and equipment of the department including the RGUMF facility for species confirmation and identification.

The authors are especially thankful to the directors of ICAR-DCFR, ICAR-NBFGR, DBT, and State Fishery Department, Government of Arunachal Pradesh for strengthening fishery research in the Department of Zoology, RGU through various institutional research programs from time to time that ultimately helped in this compilation.

The authors are grateful to Prof. Hirendra Nath Sarma, Coordinator of the "Center with Potential for Excellence in Biodiversity (CPEB-II)," Rajiv Gandhi University, Rono Hills, Doimukh and the former Coordinator Prof. A.K. Das (CPEB-I) for their continuous and untiring necessary support under the biodiversity program of Rajiv Gandhi University, Doimukh. The authors also gratefully acknowledge the support and encouragement provided by Prof. Rajiv K. Singh, Coordinator of Bioinformatics Centre for designing web pages and database of the identified fish species of the Rajiv Gandhi University Museum of Fishes (RGUMF), Doimukh.

The authors express their sincere gratitude to Prof. Waikhom Vishwanath, Department of Life Sciences, Manipur University and IUCN-South Asia Regional Chair, Freshwater Fish Specialist Group, for sharing taxonomic knowledge and skill as well as his boundless permission to access the specimens from the Manipur University Museum of Fishes, Canchipur along with making available of valuable literature and the photograph in Fig. 139.

The authors owe special thanks to the Dr. Bikramjit Sinha and Dr. Shantabala Devi Gurumayum, Arunachal Pradesh Regional Centre-Zoological Survey of India (ZSI), Itanagar for giving us access to the type and nontype specimens in ZSI museum, Itanagar. Further, the authors acknowledge Dr. Gurumayum for providing the photograph in Fig. 219.

The authors also thank Mr. Lakpa Tamang and Dr. Bikram Bakalial for providing valuable photographs of some indigenous species of Arunachal Pradesh.

Authors are obliged to the contribution of Dr. Kenjum Bagra, Dr. Boni Amin Laskar, Mr. Millo Ruja, Dr. Kento Kado, Dr. Mrinalini Choudhury, and Dr. K. Nebeshwar Sharma for initiating the process of founding RGUMF—an eastern Himalayan fish repository without which this work could not be materialized.

In the context of enriching the resources of RGUMF, the authors are highly pleased to acknowledge the financial support from various projects and schemes sanctioned by UGC (through CPEB-I and CPEB-II, F.No. 1-15/2002/2011, NS/PE, dated March 20, 2012), DBT, ICAR, and the State Directorate of Fishery, Government of Arunachal Pradesh.

The authors also thank their coworkers viz. Dr. Arijit Ganguly, Dr. Vivekananda Safi, Dr. Akash Kachari, Dr. Budhin Gogoi, Mrs. Rashmi Dutt, Ms. Rajarshee Saikia, Ms. Okyam Ering, Mr. Ram Kumar, Mr. H. Vivek, Mr. Kangkim Ratan, Ms. Awarlin Chetia, Ms. Krema Queen, Mr. Phurin Senthong, Ms. Renu Loyi, Ms Amo Mihin, Mr. Mohsin Raja, and Mr. Teriat Matey for their help in collection, identification, and discussion during the preparation of this book.

Contributions of Mr. Arun Kumar Gogoi (Laboratory Assistant, CPEB-I), Mr. Dulal Biswas, Mr. Pabitra Basumatary, Mr. Dhanabir Tamang, Tok Loma and Pradeep Pegu for collection of various species from the fields are gratefully acknowledged.

The authors like to extend their gratitude to Prof. Otem Padung, Department of Commerce, RGU; Dr. David Pertin, Deputy Registrar, RGU; Dayum Davi, District Fishery Development Officer, Yupia, Directorate of Fisheries, Arunachal Pradesh; Mr. Taro Chatung, Accredited Journalists of Electronic Media, Itanagar; Mr. Tichi Rimpum, local resident of Emchi Village, Doimukh, Mr. Leki Wangchuk and Miss Gipe, student of Department of Zoology, RGU for providing the vernacular names of endemic fishes of Arunachal Pradesh. The authors are also thankful to Mr. M. Stelin Singh, Assistant Professor, RGU, for his suggestion and donating fish specimens collected from Tawang and West Kameng district. Last but not least, authors are indebted to Prof. Nihar Ranjan Chattopadhyay of West Bengal University of Animal and Fisheris Sciences, West Bengal, for his generous advice and encouragement during compilation of this book.

Chapter 1

Overview

There are about 32,000 species of freshwater and marine fishes belonging to 536 families and 85 orders, and about 43% (13,760 species) of them are predominantly freshwater fishes (Nelson et al., 2016). The number of fish species that live in freshwater (less than 1% of all water on earth) is disproportionately larger compared to marine (97% of all water on Earth) habitats. The understanding of the diversity of ichthyofauna continues to grow rapidly, and many new discoveries are being made in the Southeast Asian regions including the northeastern region of India.

Biogeographically, the hill state Arunachal Pradesh in northeast India (Fig. 1) lies in between 26.28°N and 29.30°N latitude and 91.20°E and 97.30°E longitude, covering an area of 83,743 km^2. The state is located in the eastern part of the Himalaya biodiversity hot spot. The state is also one of the regions under the freshwater biodiversity hot spot of the world (Kottelat and Whitten, 1996). The forest coverage area of the state is 80.93% and is still intact due to low human density. The region is blessed with a unique drainage system of the upper Brahmaputra River consisting of a network of about 11 major rivers namely, Dikrong, Siang/Dihang, Ranga, Lohit, Noadihing, Kameng, Subansiri, Tawang Chu, Kamala, Dibang, Tirap, and countless small tributaries of these river systems (Figs. 2–8). The state also possesses number of lentic habitat (Fig. 9), viz. lakes, reservoirs, natural ponds, and tanks along with rice wetlands distributed in its different altitudes, viz. low altitude (up to 300 msl), mid-altitude (301–1200 msl), and high altitude (1201 msl and above). The state being situated in the high rainfall zone provides habitats for diverse aquatic life forms within its lentic and lotic environment. Therefore, from the time immemorial the native people of the state are indispensable with various aquatic resources specially fishes for their traditional rituals and ceremony (Monpa tribe); food as well as livelihood. The utilization patterns of such resources confirm that fishing is a common practice of the people for their protein diet and all the tribes are habituated in capture fishery in the streams and rivers. However, scientific explorations of ichthyofaunal composition of many water bodies of the state are still pending due to poor road connectivity, dense forest covers with wild animals, steep mountain terrains, and language barriers with different inhabitant tribes.

Pioneering works on systematic account of the fishes of the state were accomplished by McClelland (1839), Chaudhuri (1913), Hora (1921a,b), Jayaram and Mazumdar (1964), Menon (1964), Srivastava (1966), Dutta Choudhary

Biodiversity of Fishes in Arunachal Himalaya. https://doi.org/10.1016/B978-0-12-815556-1.00001-7

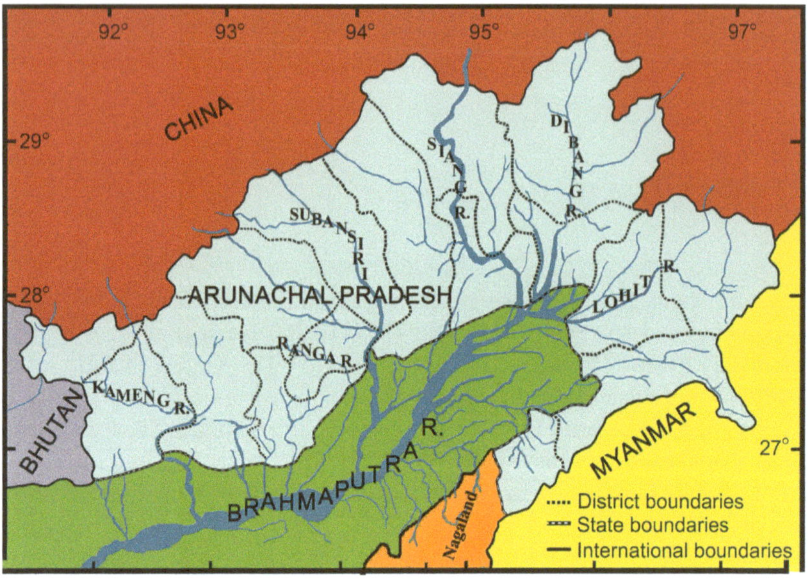

FIG. 1 Map of Arunachal Pradesh state.

FIG. 2 A view of the Siang River from the Ranaghat Bridge at Pasighat, East Siang District.

and Sen (1977), and Barman (1985). For the first time, Nath and Dey (2000) attempted to document the ichthyofauna with a record of 131 species of fish from the state. Further, Bagra et al. (2009) published an article citing a total of 213 species of fish from the state. However, accurate account of the fish taxa of the state has never been determined because of regular discovery of

FIG. 3 A portion of the Sisiri River at Lower Dibang valley district.

FIG. 4 A view of the Lohit River at the site of Alubari Ghat Bridge in Lohit district.

FIG. 5 A view of the Sinking River at Anpum in Lower Dibang Valley district.

FIG. 6 A view of the Kameng River at Tippi in West Kameng district.

FIG. 7 A view of Yomgo River at Khamba in West Siang district.

new species, resurrection, synonymization, and new occurrences regularly. Further, species discovered from different unconnected drainages of other basins have also been reported from the state by many workers, for example, *Glyptothorax trilineatus*, *Olyra horae*, *Exostoma berdmorei*, and *Tariqilabeo burmanicus* from Chindwin basin of Myanmar; *Exostoma stuarti* from Nam Yak River of Myanmar, *Schistura sikmaiensis*, *Schistura kangjupkhulensis*, and *Schistura manipurensis* from Chindwin basin in Manipur, India; *Channa orientalis* from East Indies, *Clarias batrachus* from Java, *Pterocryptis torrentis* from Lapae stream, Thailand, *Mystus montanus* from the Peninsular India, etc.

FIG. 8 Tenga River at Tenga valley in West Kameng district.

FIG. 9 A view of Sela Lake at Sela Pass in Tawang district.

Probably, misidentification of native species with similar looking species from other basins has created controversy. In fact, the distributions of hill stream fishes are restricted to a particular region and they are not likely to occur outside of that particular basin (Ng and Rachmatika, 2005).

In the two last decades, many ichthyologists showed keen interest in ichthyofaunal exploration of the state and discovered many new fish taxa. The notable contributors on fish systematics of the state are Tamang and Chaudhry (2011), Vishwanath and Darshan (2007), Ng and Tamang (2012), Lokeshwor and Vishwanath (2012), Tamang (2013), Sinha and Tamang (2014), Tamang and Sinha (2014), Tamang and Sinha (2014), Arunachalam et al. (2014),

Darshan et al. (2014, 2015, 2016), Kosygin (2012), Tamang et al. (2015), Arunkumar (2016), Das and Darshan (2017), etc.

The threats to the fishery resources of the state are mainly overfishing using dynamite, electro-fisher, poison, etc.; habitat destruction by the removal of riverbed substrate, for quarry work by damming, releasing pollutants, etc.; several anthropogenic factors related to urbanization road construction, forest clearance, etc. Besides, all these factors have direct impact on the pattern of distribution of most of the endemic fish taxa of the state being restricted to a particular areas having tremendous pressure on their population attributes. The species that are included in the IUCN endangered categories also exposed gradually to more critical conditions due to cumulative effect of the above-mentioned interference in the river systems of the state. So, considering the situation, it has been concluded that Himalayan drainages system of the state are gradually altered and inhabitation of piscatorial resources dwindled regularly due to threats from climate change, natural calamities, or because of anthropogenic interferences.

Further, many recent changes in the fish systematics have occurred in many fish taxa endemic to this mountainous state. The family Ritidae has been erected to accommodate the genera Rita and Nanobagrus. The earlier Schilbid catfishes of Southeast Asia have been now placed in the family Ailiidae. Many taxon formerly known under the order Perciformes such as *Anabas*, *Channa*, *Badis*, *Chanda*, and *Parambassis* are now placed in family Anabantiformes. All these major changes in the fish systematics of the region have been reflected in this compilation.

With the above backdrop, this book aimed to catalogue the ichthyofaunal resources of the state, their identity details, and classification based on the description of the registered topotype specimens of RGUMF and ZSI, Itanagar and type specimens of all the recently discovered fish species from the state. Further, this book is updated by incorporating all the major changes that recently took place in the area of fish systematic and nomenclatural sciences.

Chapter 2

Methodologies

Fishes were sampled using various types of indigenous fishing techniques and fishing contraception like nets, traps, electrofishing (Samus-725G), etc., and were also received from fishermen encountered at the sampling sites and local fish markets. The collected samples were preserved in 5%–10% formaldehyde solution after euthanization with MS-222 following the rules of Rajiv Gandhi University Institutional Animal Ethical Committee, Doimukh. The samples were brought to the laboratory and sorted out and identified upto species level. All the species samples were deposited in the Rajiv Gandhi University Museum of Fishes (RGUMF), Doimukh by providing a museum registration number along with the collection data. Museum fish specimens (holotype, paratype, topotype) related to Arunachal Pradesh, which are preserved in the Arunachal Pradesh Regional Centre Zoological Survey of India (APRC-ZSI), Itanagar; the Manipur University Museum of Fishes (MUMF), Canchipur, Manipur; and the Lakhimpur Girls' College Fish Museum (LGCFM), Lakhimpur, Assam were also examined.

Morphometric measurements were made point to point with a digital calipers to the nearest 0.1 mm. Fin rays were counted under the transmitted light using a Nikon SMZ 800 stereoscopic microscope. Identification up to generic level was done following Jayaram (2006) and Vishwanath et al. (2007). Presently adopted genera specific counts and measurements methodologies were followed as per Roberts (1983) (for *Bagarius*), Kottelat (1990) (Nemacheiline), Kullander and Britz (2002) (*Badis*), Zhou et al. (2011) (*Creteuchiloglansis*), Ng and Dodson (1999) (*Glyptothorax*, *Mystus*, and *Sperata*); Ng and Kottelat (2001) (*Batasio*), Ng and Wright (2009) (*Amblyceps*), Musikasinthorn (1998) (*Channa*), Nebeshwar and Vishwanath (2013) (*Garra*), Pethiyagoda et al. (2012) (*Puntius*, *Pethia*, and *Systomus*); Ng and Kottelat (1998) (*Pseudolaguvia*), Kottelat (2003) (*Parambassis*, *Chanda*); and Bornbusch (1991) (*Ompok* and *Pterocryptis*). Recent changes in fish systematics and classification; and taxa level sequence of arrangement in the book followed Nelson et al. (2016).

The method for clearing and staining of bones follows Hollister (1934). Nomenclature of bones were followed as per Greenwood (1976) and vertebral counts as per Roberts (2000).

The photographs of the fresh specimens were taken at the collection site whenever possible to document the original or the true color of the fish specimens. The vernacular names of the fishes were recorded from the local fishermen and native inhabitant communities.

Biodiversity of Fishes in Arunachal Himalaya. https://doi.org/10.1016/B978-0-12-815556-1.00002-9

Chapter 3

Systematic Index

Order Suborder	Family	Genus	Species
I. ORDER: ANGUILLIFORMES			
	1. Anguillidae	*Anguilla* Schrank, 1798	*Anguilla bengalensis* (Gray)
II. ORDER: OSTEOGLOSSIFORMES (bony tongues)			
	2. Notopteridae	*Notopterus* Lacepède, 1800	*Notopterus notopterus* (Pallas)
		Chitala Fowler, 1934	*Chitala chitala* (Hamilton)
III. ORDER: CLUPEIFORMES (Herrings)			
	3. Clupeidae	*Gudusia* Fowler, 1911	*Gudusia chapra* (Hamilton)
IV. ORDER: CYPRINIFORMES			
	SUPERFAMILY: CYPRINOIDEA **4. Cyprinidae** (minnows, carps, loaches)		
	Subfamily: Danioninae	*Amblypharyngodon* Bleeker, 1860	*Amblypharyngodon mola* (Hamilton)
		Cabdio Hamilton, 1822	*Cabdio jaya* (Hamilton)
			Cabdio morar (Hamilton)
		Opsarius McClelland, 1838	*Opsarius barna* (Hamilton)
			Opsarius tileo (Hamilton)

Continued

Biodiversity of Fishes in Arunachal Himalaya. https://doi.org/10.1016/B978-0-12-815556-1.00003-0

Order Suborder	Family	Genus	Species
		Barilius Hamilton, 1822	*Barilius bendelisis* (Hamilton)
			Barilius vagra (Hamilton)
			Barilius barila (Hamilton)
			Barilius shacra (Hamilton)
		Laubuka Bleeker, 1859	*Laubuka laubuca* (Hamilton)
		Chela Hamilton, 1822	*Chela cachius* (Hamilton)
		Danio Hamilton, 1822	*Danio assamila* Kullander
			Danio rerio (Hamilton)
		Devario Heckel, 1843	*Devario aequipinnatus* (McClelland)
			Devario devario (Hamilton)
			Devario assamensis (Barman)
		Esomus Swainson, 1839	*Esomus danrica* (Hamilton)
		Raiamas Jordan, 1919	*Raiamas bola* (Hamilton)
		Rasbora Bleeker, 1859	*Rasbora daniconius* (Hamilton)
			Rasbora rasbora (Hamilton)
		Salmostoma Swainson, 1839	*Salmostoma bacaila* (Hamilton 1822)
			Salmostoma phulo (Hamilton)
		Securicula Günther, 1868	*Securicula gora* (Hamilton)

Order Suborder	Family	Genus	Species
	Subfamily: Labeoninae	*Bangana* Hamilton, 1822	*Bangana dero* (Hamilton)
		Gibelion Heckel 1843	*Gibelion catla* (Hamilton)
		Cirrhinus Oken, 1817	*Cirrhinus mrigala* (Hamilton)
			Cirrhinus reba (Hamilton)
		Tariqilabeo Mirza & Saboohi, 1990	*Tariqilabeo latius* (Hamilton)
		Garra Hamilton, 1822	*Garra annandalei* Hora
			Garra lissorhynchus (McClelland)
			Garra kalpangi Nebeshwar et al.
			Garra lamta (Hamilton)
			Garra kempi Hora
			Garra rupicola (McClelland)
			Garra magnidiscus Tamang
			Garra arupi Nebeshwar et al.
			Garra arunachalensis Nebeshwar & Vishwanath
			Garra birostris Nebeshwar & Vishwanath
			Garra tamangi Gurumayum & Kosygin
			Garra minima Arunachalam et al.

Continued

Order Suborder	Family	Genus	Species
			Garra alticaputus Arunachalam et al.
			Garra nigricauda Arunachalam et al.
			Garra kimini Arunachalam et al.
		Labeo Cuvier, 1816	*Labeo calbasu* (Hamilton)
			Labeo gonius (Hamilton)
			Labeo pangusia (Hamilton)
			Labeo rohita (Hamilton)
			Labeo bata (Hamilton)
			Labeo boga (Hamilton)
			Labeo dyocheilus (McClelland)
	Subfamily: Cyprininae	*Barbonymus* Kottelat 1999	*Barbonymus gonionotus* (Bleeker)
		Cyprinus Linnaeus, 1758	*Cyprinus carpio* Linnaeus
		Carassius Jarocki, 1822	*Carassius carassius* (Linnaeus)
		Chagunius Smith, 1938	*Chagunius chagunio* (Hamilton)
		Neolissochilus Rainboth, 1985	*Neolissochilus hexagonolepis* (McClelland)
			Neolissochilus hexastichus (McClelland)
		Osteobrama Heckel, 1843	*Osteobrama cotio* (Hamilton)

Order Suborder	Family	Genus	Species
		Systomus McClelland, 1838	*Systomus immaculatus* McClelland
			Systomus sarana (Hamilton)
			Systomus clavatus (McClelland)
		Oreichthys Smith, 1933	*Oreichthys crenuchoides* Schafer
		Puntius Hamilton, 1822	*Puntius sophore* (Hamilton)
			Puntius chola (Hamilton)
		Pethia Pethiyagoda, Meegaskumbura & Maduwage, 2012	*Pethia conchonius* (Hamilton)
			Pethia ticto (Hamilton)
			Pethia guganio (Hamilton)
			Pethia gelius (Hamilton)
		Schizothorax Heckel, 1838	*Schizothorax richardsonii* (Gray)
			Schizothorax progastus (McClelland)
		Semiplotus Bleeker, 1860	*Semiplotus semiplotus* (McClelland)
		Tor Gray, 1834	*Tor putitora* (Hamilton)
			Tor tor (Hamilton)
	Subfamily: Squaliobarbinae	*Ctenopharyngodon* Steindachner, 1866	*Ctenopharyngodon idella* (Valenciennes)

Continued

Order Suborder	Family	Genus	Species
	Subfamily: Xenocyprinae	*Hypophthalmichthys* Bleeker 1860	*Hypophthalmichthys molitrix* (Valenciennes)
			Hypophthalmichthys nobilis (Richardson)
	5. Psilorhynchidae (mountain carps)	*Psilorhynchus* McClelland, 1838	*Psilorhynchus balitora* (Hamilton)
			Psilorhynchus homaloptera Hora & Mukerji
			Psilorhynchus arunachalensis (Nebeshwar, Bagra and Das)
			Psilorhynchus sucatio (Hamilton)
	SUPER FAMILY: COBITOIDEA	*Botia* Gray, 1831	*Botia dario* (Hamilton)
	6. Botiidae (botiid loaches)		*Botia rostrata* Gunther
	7. Cobitidae	*Canthophrys* Swainson, 1838	*Canthophrys gongota* (Hamilton)
		Lepidocephalichthys Bleeker, 1863	*Lepidocephalichthys guntea* (Hamilton)
			Lepidocephalichthys annandalei Chaudhuri
			Lepidocephalichthys arunachalensis (Datta & Barman)
			Lepidocephalichthys goalparensis Pillai & Yazdani
		Pangio Blyth 1860	*Pangio pangia* (Hamilton)
	8. Balitoridae	*Balitora* Gray, 1830	*Balitora brucei* Gray

Order Suborder	Family	Genus	Species
	9. Nemacheilidae (Stone loaches)	*Aborichthys* Chaudhuri, 1913	*Aborichthys boutanensis* (McClelland)
			Aborichthys elongatus Hora
			Aborichthys tikaderi Barman
			Aborichthys waikhomi Kosygin
			Aborichthys sp.
		Acanthocobitis Peters, 1861	*Acanthocobitis pavonaceous* (McClelland)
		Paracanthocobitis Grant, 2007	*Paracanthocobitis botia* (Hamilton)
		Schistura McClelland, 1838	*Schistura walongensis* (Tamang & Sinha)
			Schistura beavani (Gunther)
			Schistura zonata McClelland
			Schistura devdevi (Hora)
			Schistura savona (Hamilton)
			Schistura scaturigina (McClelland)
			Schistura multifasciata (Day)
			Schistura tirapensis Kottelat
			Schistura rosammae (Sen)
			Schistura sp 1
			Schistura sp 2

Continued

Order **Suborder**	**Family**	**Genus**	**Species**
		Physoschistura Bănărescu & Nalbant, 1982	*Physoschistura dikrongensis* Lokeshwar & Vishwanath
			Physoschistura harkishorei Das & Darshan
			Physoschistura sp.
V. ORDER: SILURIFORMES			
SUBORDER: SILUROIDEA	**10. Siluridae** (sheatfishes)	*Ompok* Lacepède, 1803	*Ompok bimaculatus* (Bloch)
			Ompok pabo (Hamilton)
			Ompok pabda (Hamilton)
		Pterocryptis Peters, 1861	*Pterocryptis gangelica* Peters
			Pterocryptis indicus (Datta et al.)
		Wallago Bleeker, 1851	*Wallago attu* (Bloch & Schneider)
	11. Chacidae (Squarehead, angler, or frogmouth catfishes)	*Chaca* Gray, 1831	*Chaca chaca* (Hamilton)
	12. Ritidae (retas and nanobagrids)	*Rita* Bleeker, 1853	*Rita rita* (Hamilton)
	13. Ailiidae (Asian schilbeids)	*Ailia* Gray, 1830	*Ailia coila* (Hamilton)
		Clupisoma Swainson, 1838	*Clupisoma garua* (Hamilton)
			Clupisoma montanum (Hora)
		Eutropiichthys Bleeker, 1862	*Eutropiichthys vacha* (Hamilton)
	14. Horabagridae (imperial or sun catfishes)	*Pachypterus* Swainson, 1838	*Pachypterus atherinoides* (Bloch)

Order Suborder	Family	Genus	Species
	15. Bagridae (bagrid catfishes)	*Batasio* Blyth, 1860	*Batasio fasciolatus* Ng
			Batasio merianiensis Chaudhuri
			Batasio batasio (Hamilton)
			Batasio tengana (Hamilton)
			Batasio spilurus Ng
		Chandramara Jayaram, 1972	*Chandramara chandramara* (Hamilton)
		Mystus Scopoli, 1777	*Mystus cavasius* (Hamilton)
			Mystus bleekeri (Day)
			Mystus tengara (Hamilton)
			Mystus carcio (Hamilton)
			Mystus dibrugarensis (Chaudhuri)
		Sperata Holly, 1939	*Sperata aor* (Hamilton)
			Sperata seenghala (Sykes)
		Olyra McClelland, 1842	*Olyra longicaudata* McClelland
			Olyra praestigiosa Ng & Ferraris

Continued

Order Suborder	Family	Genus	Species
	SUPERFAMILY: SISOROIDEA		
	16. Amblycipitidae (torrent catfishes)	*Amblyceps* Blyth, 1858	*Amblyceps arunachalensis* Nath & Dey
			Amblyceps apangi Nath & Dey
			Amblyceps laticeps (McClelland)
			Amblyceps waikhomi Darshan et al.
			Amblyceps mangois (Hamilton)
	17. Sisoridae (sisorid catfishes)		
	Subfamily: Sisorinae	*Sisor* Hamilton, 1822	*Sisor rhabdophorus* Hamilton
		Bagarius Bleeker, 1853	*Bagarius bagarius* (Hamilton)
		Gagata Bleeker, 1858	*Gagata cenia* (Hamilton)
		Gogangra Roberts, 2001	*Gogangra viridescens* (Hamilton)
	Subfamily: Glyptosterninae	*Creteuchiloglanis* Zhou, Li & Thomson, 2011	*Creteuchiloglanis kamengensis* (Jayaram)
			Creteuchiloglanis payjab Darshan et al.
			Creteuchiloglanis arunachalensis Sinha & Tamang
		Exostoma Blyth, 1860	*Exostoma labiatum* (McClelland)
			Exostoma tenuicaudata Tamang et al.

Order Suborder	Family	Genus	Species
		Glyptothorax Blyth, 1860	*Glyptothorax cavia* (Hamilton)
			Glyptothorax telchitta (Hamilton)
			Glyptothorax botius (Hamilton)
			Glyptothorax indicus Talwar
			Glyptothorax dikrongensis Tamang & Chaudhry
			Glyptothorax striatus (McClelland)
			Glyptothorax pantherinus Anganthoibi & Vishwanath
			Glyptothorax mibangi Darshan et al.
			Glyptothorax pasighatensis Arunkumar
		Oreoglanis Smith, 1933	*Oreoglanis majuscula* Linthoingambi & Vishwanath
			Oreoglanis pangenensis Sinha & Tamang
		Parachiloglanis Wu, He & Chu, 1981	*Parachiloglanis hodgarti* (Hora)
		Pseudecheneis Blyth, 1860	*Pseudecheneis sulcata* (McClelland)
			Pseudecheneis sirenica Vishwanath & Darshan

Continued

Order Suborder	Family	Genus	Species
	18. Erethistidae (erethistid catfishes) Subfamily: Continae	*Conta* Hora, 1950	*Conta conta* (Hamilton)
	Subfamily: Erethistinae	*Erethistoides* Hora, 1950	*Erethistoides senkhiensis* Tamang et al.
		Erethistes Müller & Troschel, 1849	*Erethistes pusillus* Müller & Troschel
			Erethistes hara (Hamilton)
			Erethistes jerdoni (Day)
		Nangra Day, 1877	*Nangra assamensis* Sen & Biswas
		Pseudolaguvia Misra, 1976	*Pseudolaguvia shawi* (Hora)
			Pseudolaguvia viriosa Ng & Tamang
			Pseudolaguvia jiyaensis Tamang & Sinha
			Pseudolaguvia magna Tamang & Sinha
	19. Clariidae (airbreathing catfishes)	*Clarias* Scopoli, 1763	*Clarias magur* (Hamilton)
			Clarias gariepinus (Burchell)
	20. Heteropneustidae (Saccobranchidae) (airsac catfishes)	*Heteropneustes* Müller, 1840	*Heteropneustes fossilis* (Bloch)
VI. ORDER: SALMONIFORMES			
	21. Salmonidae (trout, salmon, whitefish) Subfamily: Salmoninae	*Oncorhynchus* Suckley, 1861	*Oncorhynchus mykiss* (Walbaum)
		Salmo Linnaeus 1758	*Salmo trutta* Linnaeus

Order Suborder	Family	Genus	Species
VII. ORDER: GOBIIFORMES			
	22. Gobiidae (gobies)	*Glossogobius* Gill, 1859	*Glossogobius giuris* (Hamilton)
	23. Ambassidae (Chandidae) (Glass fishes)	*Chanda* Hamilton, 1822	*Chanda nama* Hamilton
		Parambassis Bleeker, 1874	*Pseudambassis ranga* (Hamilton)
			Parambassis bistigmata Geetakumari
			Parambassis baculis (Hamilton)
VIII. ORDER: CICHLIFORMES			
	24. Cichlidae (Ciclids)	*Oreochromis* Günther, 1889	*Oreochromis mossambicus* (Peters)
IX. ORDER: BELONIFORMES			
SUBORDER: BELONOIDEI	SUPER FAMILY: EXOCOETOIDEA **25. Belonidae**	*Xenentodon* Regan, 1911	*Xenentodon cancila* (Hamilton)
X. ORDER: SYNBRANCHIFORMES (swamp eels)			
SUBORDER: SYNBRAN-CHOIDEI	**26. Synbranchidae**	*Monopterus* Lacepède, 1800	*Monopterus cuchia* (Hamilton)
SUBORDER: MASTACEMBE-LOIDEI	**27. Chaudhuriidae**	*Pillaia,* Yazdani, 1972	*Pillaia indica* Yazdani
	28. Mastacembelidae	*Macrognathus* Lacepede, 1800	*Macrognathus aral* (Bloch & Schineider)
			Macrognathus pancalus Hamilton
		Mastacembelus Scopoli, 1777	*Mastacembelus armatus* (Lacepede)

Continued

Order Suborder	Family	Genus	Species
XI. ORDER: ANABANTIFORMES (Labyrinth fishes)			
SUBORDER: ANABANTOIDEI	**29. Anabantidae** (Climbing gouramis)	*Anabas* Cloquet, 1816	*Anabas testudineus* (Bloch)
	30. Osphronemidae Subfamily: Luciocephalinae	*Trichogaster* Bloch & Schneider, 1801	*Trichogaster chuna* (Hamilton)
			Trichogaster lalius (Hamilton)
			Trichogaster fasciata Bloch & Schneider
			Trichogaster labiosa Day
SUBORDER: CHANNOIDEI	**31. Channidae** (snakeheads)	*Channa* Scopoli, 1777	*Channa punctata* (Bloch)
			Channa striata (Bloch)
			Channa aurantimaculata Musikasinthorn
			Channa marulius (Hamilton)
			Channa stewartii (Playfair)
			Channa melanostigma Geetakumari & Vishwanath
			Channa bleheri Vierke
			Channa barca (Hamilton)
			Channa punctata (Bloch)
			Channa striata (Bloch)
			Channa aurantimaculata Musikasinthorn

Order Suborder	Family	Genus	Species
			Channa marulius (Hamilton)
			Channa stewartii (Playfair)
			Channa melanostigma Geetakumari & Vishwanath
			Channa bleheri Vierke
			Channa barca (Hamilton)
			Channa gachua (Hamilton)
			Channa pomanensis Gurumayum & Tamang
	32. Nandidae (Asian leaf fishes)	*Nandus* Valenciennes, 1831	*Nandus nandus* (Hamilton)
	33. Badidae (chameleon fishes)	*Badis* Bleeker, 1853	*Badis assamensis* Ahl
			Badis badis (Hamilton)
			Badis singenensis Geetakumari & Kadu
			Badis dibruensis Geetakumari & Vishwanath
XII. ORDER: SYNGNATHIFORMES			
SUBORDER: SYNGNATHOIDEI	SUPER FAMILY SYNGNATHOIDEA **34. Syngnathidae** Subfamily: Syngnathinae (pipe fishes)	*Microphis* Kaup 1853	*Microphis deocata* (Hamilton)
XIII. ORDER: TETRAODONTIFORMES			
	35. Triodontidae (three-toothed puffers)	*Leiodon* Swainson, 1839	*Leiodon cutcutia* (Hamilton)

Chapter 4

Systematic Account

I. ORDER: ANGUILLIFORMES

Body very elongate and eel-like; pelvic fin and internal basipterygium absent; pectoral fin and internal girdle may be absent, if present, located nearly midlateral in position or higher and girdle lacking connection to cranial bone; dorsal and anal fins continuous with caudal fin (caudal fin absent in some); embedded cycloid scales, or totally absent; gill openings usually narrow; gill arches elongate and gill rakers absent (except in *Protanguilla*); pyloric caeca absent; maxilla toothed, bordering mouth; two premaxillae (rarely absent), vomer (usually), and the ethmoid joined into a single bone; branchiostegal rays 6–49; swim bladder present, duct usually present; oviducts absent; opisthotic, orbitosphenoid, mesocoracoid, gular plate, posttemporal, postcleithra, supramaxilla, and extrascapular bones absent; ossified symplectic absent (one cartilaginous present in Synaphobranchidae); hyomandibular joined with quadrate; ribs present or absent.

1. Family: Anguillidae

1. Genus: *Anguilla* Schrank, 1798

Anguilla Schrank 1798: 304, 307. Type species: *Muraena Anguilla* Linnaeus 1758. Type by monotypy.

Body elongated and tube-like; heads pointed; tail compressed; pectoral fin present, usually small; cycloid scales embedded in skin; dorsal fin usually continuous with caudal and anal fins; dorsal fin inserted midway between pectoral-fin origin and vent, lateral line complete.

***Anguilla bengalensis* (Gray, 1831)**

Fig. 10.

Muraena bengalensis Gray, 1831: pl. 95, Fig. 5; type locality: the Ganges, India.

Anguilla bengalensis: Kottelat, 2013: 37 (valid).

Materials examined. RGUMF-165, 1 ex., 330 mm SL; Ranga River at Kimin, Papum Pare district, Arunachal Pradesh. RGUMF-unregistered, 1 ex.,

Biodiversity of Fishes in Arunachal Himalaya. https://doi.org/10.1016/B978-0-12-815556-1.00004-2

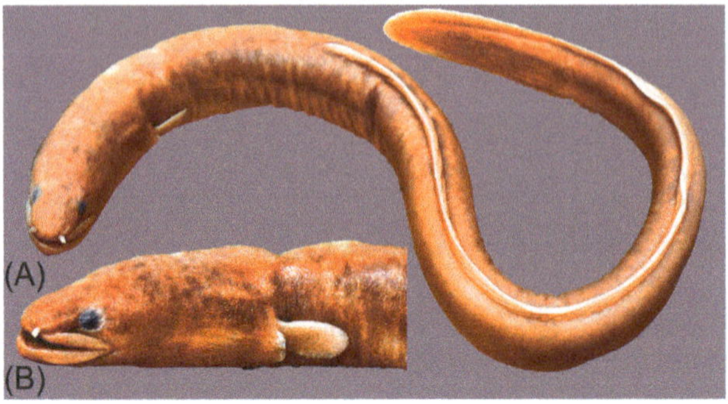

FIG. 10 *Anguilla bengalensis*, RGUMF-165, 330 mm SL: (A) dorsolateral view, and (B) lateral view of head.

420 mm SL, Dibang River at Roing, Lower Dibang Valley district, Arunachal Pradesh. RGUMF-unregistered, 2 ex., 350–418 mm SL, Dibang River at Roing, Lower Dibang Valley district, Arunachal Pradesh.

Vernacular names in Arunachal Pradesh. *Ngobe*/*Ngobeic* (in Nyishi), *Nyobi* (in Galo), and *Ngobi* (in Adi).

Diagnosis. Body elongate and tail compressed. Head: conical in dorsal view, depressed and flattened. Lips small and thick. Scales rounded, embedded in the skin. Barbels and pelvic fin absent. Dorsal and anal fins continuous with the caudal fin.

Coloration. In live specimen, body yellowish to olive brown, mottled with dark brown markings, ventrally lighter. In formalin preserved specimens, over all the body is brownish mottled with dark-brown markings.

Distribution. Widely distributed: India, Nepal, Bangladesh, Bhutan, Pakistan, Myanmar, and Sri Lanka. In Arunachal Pradesh: Known from Siang, Ranga, Kameng, Dikrong, Subansiri, and Dibang rivers.

Remarks: This species is a catadromous migratory fish, breeds in the ocean and its juvenile (elvers) migrating into estuaries and riverine freshwaters for feeding. On attaining maturity, they return to sea for breeding.

Economic importance. This species is a popular delicacy food fish.

Conservational status: Based on the information from the local village elders, the species was very abundant in Arunachal Pradesh. It was reported that, very often a shoal of this fishes use to climb up to the river bank vegetation to eat tender bamboo shoots. However, the population of this species has declined so rapidly that during the last 10 years, the species is encountered only in hook-catch occasionally. IUCN status: Near threatened (NT).

II. ORDER: OSTEOGLOSSIFORMES

Osteoglossiformes (Greek: "bony tongues") contains two suborders, the Osteoglossoidei and the Notopteroidei. A relatively primitive taxon of ray-finned fish having an intestine passing posteriorly to the left of esophagus and stomach; parasphenoid and tongue bones usually with well-developed teeth; premaxilla small and fixed to the skull; supramaxilla absent; caudal fin skeleton with large first ural centrum and with no urodermals, epurals may be one or more and they are fused with uroneurals (except in *Hiodon*); caudal skeleton bears 16 or fewer branched rays; nasal capsule rigid, no antorbital-supraorbital system; no epipleural intermuscular bones; one (in *Hiodon* and *Pantodon*) or two pyloric (in other Osteoglossiformes).

2. Family: Notopteridae

2. Genus: *Notopterus* Lacepède, 1800

Notopterus Lacepède 1800: 189. Type species: *Gymnotus notopterus* Pallas, 1769: 40, by absolute tautonymy, in the synonymy of *Gymnotus kapirat* Lacepède, 1800: 190.

Body highly compressed, oblong and tapers toward caudal fin; cranio-dorsal profile straight or slightly concave; small dorsal fin, inserted midway between snout tip and base of caudal fin; anal fin long; caudal fin small and fused with anal fin; abdomen edge anterior to pelvic fin keeled with double serration; well-developed teeth on tongues and jaws; barbels absent, a small appendage on each nostril; body scales very small.

***Notopterus notopterus* (Pallas, 1769)**

Fig. 11.

FIG. 11 *Notopterus notopterus*, RGUMF-163, 135 mm SL: lateral view.

Gymnotus notopterus Pallas, 1769: 40, pl. 6, Fig. 2; type locality: Ambon, Indonesia.

Notopterus notopterus: Kottelat, 2013: 32 (valid).

Material examined. RGUMF-163, 4 exs., 115–180 mm SL; Ranga River at Kimin, Papum Pare district, Arunachal Pradesh.

Vernacular names in Arunachal Pradesh. *Kanduli* (in Nyishi, Adi, & Galo).

Diagnosis. Body strongly compressed, oblong and tapers toward the tail; abdomen keeled with pre-pelvic double serration; preorbital bone serrated; dorsal fin base short, inserted more or less in the middle of body length (standard length); pelvic fin rudimentary; anal fin originated immediately behind pelvic fin and continuous with caudal fin.

Coloration. Silvery in live specimens, no transverse bars on back. Body color varies from silvery to light grayish with regard to habitats. In 10% formalin, over all the body is brownish.

Distribution. Widely distributed in India, Nepal, Bangladesh, Pakistan, Indonesia, Myanmar, and Sri Lanka. In Arunachal Pradesh: known from all food hill water bodies.

Remarks. *N. notopterus* is very similar in appearance with *Chitala chitala* but can be easily distinguished from the latter by its absence (vs. presence) of transverse bars on back, and slightly concave (vs. strongly concave) cranio-dorsal profile.

Economic importance. This species is a popular delicacy food fish.

Conservational status: This widespread species is abundantly encountered in all catches from the low-lying riverine and stagnant water bodies. IUCN status: Least concern (LC).

3. Genus: *Chitala* Fowler, 1934

Chitala (subgenus of *Notopterus*) Fowler 1934: 244. Type species: *Mystus chitala* Hamilton 1822, type by original designation (also by absolute tautonymy).

Body highly compressed, oblong, tapers toward caudal fin; cranio-dorsal profile distinctly concave; abdomen edge keeled with scutes up to pelvic fin insertion; scales small or minute; dorsal fin short, inserted midway between the snout tip and the base of caudal fin; pelvic fin rudimentary, anal fin long and confluent with caudal fin. A series of golden or silvery bars along the back, resulting in a faint striped appearance; sometimes indistinct, non-ocellated dark spots toward the far rear of the body.

***Chitala chitala* (Hamilton 1822)**

Fig. 12.

FIG. 12 *Chitala chitala*, RGUMF-unregistered, 608 mm SL: lateral view.

Mystus chitala Hamilton, 1822: 236, 382, type locality: Rivers of Bengal and Behar [Bihar], India.

Chitala chitala: Kottelat, 2013: 31 (valid).

Material examined. RGUMF-unregistered, 608 mm SL, purchase from Doimukh Bazar, Papum Pare district, Arunachal Pradesh.

Vernacular names in Arunachal Pradesh. *Chital* (in Nyishi, Adi, & Galo).

Diagnosis. Adult size large, reaching up to 1220 mm SL; cranio-dorsal profile strongly concave; presence of a series of 14–15 golden to silvery bars along the back and another series of small 2–4 non-ocellated dark spots sometimes appear in front of caudal fin above anal fin.

Coloration. Body silvery laterally and light grayish dorsally, 14–15 silvery transverse bars on the back.

Distribution. Widely distributed in the Indian subcontinent: India, Nepal, Bangladesh, and Pakistan. In Arunachal Pradesh: known only from food hill areas.

Remarks. This species is very similar to *C. chitala* (for distinguishing character see above). *C. chitala* is a species larger than *N. notopterus* reaching up to ca. 1220 mm TL (vs. 190 mm SL in *N. notopterus*). The species frequently escape by cutting the net with its sharp osseous keeled abdomen and even hurt the fisherman during fishing.

Economic importance. This species is a highly prized and popular delicacy food fish.

Conservational status: Wild populations have been declining fast throughout its range of distribution due to overharvesting. Induced breeding has been successfully undertaken in many fishery farms in the states of Assam and West Bengal, India for seed production and culture. Regular wrenching program can replenish the wild population. IUCN status: Near threatened (NT).

III. ORDER: CLUPEIFORMES

Body generally silvery in color, spindle shaped; usually have abdominal or pelvic scutes; recessus lateralis present; lacking parasphenoid teeth; no large foramen on the anterior ceratohyal; parietals separated by the supraoccipital; no lateral line; gill rakers long and numerous, and most are plankton feeders.

3. Family: Clupeidae

4. Genus: *Gudusia* Fowler, 1911

Gudusia (subgenus of *Sardinella*) Fowler 1911: 207. Type species: *Clupanodon chapra* Hamilton 1822. Type by original designation (also monotypic).

Body fairly deep and compressed, fusiform shaped; abdominal edge keeled and bears a row of scutes; upper jaw with notch for lower jaw occlusion; scales small, irregular except on upper flank.

***Gudusia chapra* (Hamilton 1822)**
Fig. 13.

FIG. 13 *Gudusia chapra*, RGUMF-164, 76.3 mm SL: lateral view.

Clupanodon chapra Hamilton 1822. 248, 383; type locality: Ganges River in Northern Bihar. *Gudusia chapra*: Menon, 1999: 7 (valid).

Material examined. RGUMF-164, 5 exs., 70.0–100 mm SL, Dikrong River at Doimukh, Papum Pare district, Arunachal Pradesh.

Vernacular names in Arunachal Pradesh. *Chapila* (in Nyishi, Adi, & Galo).

Diagnosis. Body deep, fusiform, and laterally compressed, abdomen with a row of 28–29 scutes. A single triangular pectoral axillary scale and anal fin with 18–22 rays.

Coloration. Body brownish on back, flank silvery. A blackish blotch appears behind the gill opening.

Distribution. The Ganges and Brahmaputra systems in India, Bangladesh, and Nepal, and also known from the Mahanadi River, Orissa. In Arunachal Pradesh: known only from food hill areas.

Remarks. *Gudusia variegata* (Day, 1870) known from the Irrawaddy River system of Myanmar is the only congener of *G. chapra*.

Economic importance. This fish has both food and ornamental value.

Conservational status: This species is abundantly encountered in all the low-land water bodies in the Ganga-Brahmaputra river basin. IUCN status: Least concern (LC).

IV. ORDER: CYPRINIFORMES

Head scaleless; no bony plates on body; upper jaw usually protractile jaws and palate of mouth always toothless; branchiostegal rays three; spine-like rays in dorsal fin of some species; adipose fin absent (except in some cobitoids); kinethmoid present; palatine articulating in a socket of the endoptrygoid; fifth ceratobranchial (the pharyngeal bone) enlarged, with teeth ankylosed to the bone (pharyngeal teeth absent in gyrinocheilids); pharyngeal teeth opposite to enlarged posterior process of basioccipital bone rather than to upper pharyngeal elements, the basioccipital process against which the pharyngeal teeth press usually covered by a pad; elongation of the lateral process of the second vertebral centrum; projecting into somatic musculature; ascending process to premaxillae.

4. Family: Cyprinidae

Subfamily: Danioninae

5. Genus: *Amblypharyngodon* Bleeker, 1860

Amblypharyngodon Bleeker 1860: 433. Type species: *Cyprinus mola* Hamilton 1822. Type by being a replacement name.

Body elongated and compressed; abdomen rounded; eyes large, upper lip absent, no rostral cap and no horny jaw sheaths; a notch on upper jaw which fits the symphyseal knob of lower jaw; no barbels; dorsal fin inserted distinctly behind pelvic fins; scales small; lateral line incomplete; and caudal fin forked.

***Amblypharyngodon mola* (Hamilton 1822)**

Fig. 14.

FIG. 14 *Amblypharyngodon mola*, RGUMF-28, 57.0 mm SL: lateral view.

Cyprinus mola Hamilton, 1822: 334, 392, pl. 38, Fig. 92; type locality: Gangetic provinces, India.

Amblypharyngodon mola: Kottelat, 2013: 73 (valid).

Material examined. RGUMF-28, 4 exs., 54.0–57.0 mm SL, Balijan, Papum Pare district, Arunachal Pradesh.

Vernacular names in Arunachal Pradesh. *Morula* (in Nyishi, Adi, & Galo).

Diagnosis. Body elongated, highly compressed, abdomen rounded, barbels absent. Lower jaw prominently longer than the upper jaw with a symphyseal knob which fits into the notch of upper jaw. Lateral line incomplete with 12 pierced scales and 66 in the row. Eyes large, scales minute.

Coloration. Body silvery to golden yellow with a faint band on the side.

Distribution. Widely distributed: India, Nepal, Bangladesh, and Pakistan (Indus basin). In Arunachal Pradesh: known only from food hill areas.

Remarks. Records of *A. mola* from the Chindwin basin in Manipur (Northeast India) and Myanmar need verification.

Economic importance. This fish has both food and ornamental value.

Conservational status: This fish is a widespread species occurring in three neighboring river basins: the Ganga-Brahmaputra, Chindwin-Irrawaddy, and Indus river systems. The fish is found in all lowland slow moving rivers and stagnant water bodies of these basins. It breeds naturally even in small ponds and streams. IUCN status: Least concern (LC).

6. Genus: *Cabdio* Hamilton, 1822

Cabdio (subgenus of *Cyprinus*) Hamilton 1822: 333, 392. Type species: *Cyprinus* (*Cabdio*) *jaya* Hamilton 1822. Type by subsequent designation.

Body elongated and moderately compressed; rounded abdomen; suborbital process well developed; mouth inferior and small; no lips and barbels; dorsal fin inserted above or behind pelvic fins, lack any osseous rays; lateral line decurved and complete, running in lower half of caudal peduncle; scales moderate in size.

Key to species of *Cabdio*:

1a. 7–8 anal-fin branched rays, 52–60 lateral line scales.........................
.... *C. jaya*

1b. 9–10 anal-fin branched rays, 38–42 lateral line scales.............................
.....................*C. morar*

***Cabdio jaya* (Hamilton 1822)**
Fig. 15.

FIG. 15 *Cabdio jaya*, RGUMF-11, 70.0 mm SL: lateral view.

Cyprinus jaya Hamilton, 1822: 333, 392; type locality: Northern Behar [Bihar], India.

Cabdio jaya: Eschmeyer, 2017 (valid).

Material examined. RGUMF-10–11, 3 exs., 57.0–70.0 mm SL, streams at Balijan and Doimukh, Papum Pare district, Arunachal Pradesh.

Vernacular names in Arunachal Pradesh. *Ngopii-yalik* (in Nyshi); *Ngopii* (in Adi).

Diagnosis. Pectoral and pelvic fin with small axillary scales. Lateral line with 52–60 scales, anal fin with ii–iii,7–8 rays; 2 rows of pharyngeal teeth, 8 transverse scale rows between dorsal fin and lateral line and 6½–7 between lateral line and pelvic-fin origin.

Coloration. Head and body sides silvery; black tinge on the back.

Distribution. Known from the Ganga-Brahmaputra river system in India, Nepal, and Bangladesh. In Arunachal Pradesh: the fish is occurring in the mid-altitude to foot hill waters of many rivers, viz., Kameng, Subansiri, Dikrong, Ranga, Siang, Lohit, Buridhing, and Noadhing (Nath and Dey, 2000) (additional range of distribution see material examined).

Remarks. *C. jaya* differs from *C. morar* in having fewer (7–8 vs. 9–10) anal-fin branched rays and more (52–60 vs. 38–42) lateral line scales.

Economic importance. This fish has both food and ornamental value.

Conservational status: This species is rarely encountered during sampling from most of the rivers in Arunachal Pradesh. Fishermen and common peoples of the state treated both *C. jaya* and *C. morar* as same species. It is difficult to decide its abundancy from the information provided by the regular professional fishermen of the region. IUCN status: Least concern (LC).

***Cabdio morar* (Hamilton 1822)**

Fig. 16.

FIG. 16 *Cabdio morar*, RGUMF-330, 76.2 mm SL: lateral view.

Cyprinus morar Hamilton, 1822: 264, 384, pl. 31, Fig. 75; type locality: Yamuna and Tista Rivers, India.

C. morar: Kottelat, 2013: 84 (valid).

Materials examined. RGUMF-09, 7 exs., 53.0–96.0 mm SL, Poma River, Papum Pare district, Arunachal Pradesh. RGUMF-330, 10 exs. 65.1–78.2 mm SL, Siang River at Pasighat, East Siang district, Arunachal Pradesh, February 27, 2008.

Vernacular names in Arunachal Pradesh. *Ngopii-yalik* (in Nyshi); *Ngopii* (in Adi).

Diagnosis. Pectoral and pelvic fin with moderately large axillary scales. Lateral line scales 38–42, predorsal scales 15–18, dorsal fin with ii,7–7½ rays; anal fin with iii,9–10 rays; and 3 rows of pharyngeal teeth. Transverse scale rows from dorsal fin to above lateral line 6–7.

Coloration. Lateral sides of body and head silvery, dorsally yellowish with greenish reflection.

Distribution. Widely distributed in the Ganga-Brahmaputra river system in India, Bangladesh, and Nepal. Records of this species from Myanmar, Thailand, Iran, and Pakistan need identity verification. In Arunachal Pradesh: known only from the food hill areas.

Remarks. This species is more abundant than *C. jaya* in Arunachal Pradesh.

Economic importance. This fish has both food and ornamental value.

Conservational status: The species is widely spread in the Ganga-Brahmaputra basin. This species is very abundant and a common catch from all major rivers in Arunachal Pradesh and Assam. IUCN status: Least concern (LC).

7. Genus: *Opsarius* McClelland, 1838

Opsarius McClelland: 1838: 944. Type species: *Opsarius maculates* McClelland 1839. Type by subsequent designation.

Mouth widely cleft; body slender and usually marked with transverse green bars or spots. Dorsal small, without spines and placed behind the middle of body; anal fin long; intestinal canal very short and extending straight from the stomach to vent (McClelland, 1838).

Key to species of *Opsarius*:

1a. Barbels absent; flank with 8–11 bluish bars extending Up to lateral line; lateral line scales 37+3..*O. barna*

1b. Barbel present; 13–15 fragmented bars on flank, fragments are arranged into 3–4 vertical rows of bluish spots and blotches; lateral line scales 70–75................................*O. tileo*

***Opsarius barna* (Hamilton 1822)**
Fig. 17.

FIG. 17 *Opsarius barna*, RGUMF-12, 62.0 mm SL: lateral view.

Cyprinus barna Hamilton, 1822: 268, 384; type locality: Yamuna and Brahmaputra Rivers, India.

Opsarius barna: Kottelat, 2013: 131 (valid).

Material examined. RGUMF-12, 5 exs. 52–68 mm SL, Poma River in Papum Pare district, Arunachal Pradesh.

Vernacular names in Arunachal Pradesh. *Rebio Tapio* (in Nyishi), *Orchin* (in Galo), *Sepung* (in Adi).

Diagnosis. Barbel absent, body with widely spaced 8–11 bluish bars on flanks extending up to lateral line; dorsal fin inserted opposite to the posterior end of ventral-fin base; lateral line scales 37+3, and predorsal scales 16.

Coloration. In live specimens, body with widely spaced 8–11 bluish lateral bars extending up to lateral line, interspaces 2–3 times of bar width, a dark bar across the middle of the orbit, silvery body ground color, dorsally grayish with bluish reflection. Simple rays of dorsal fin blackish.

Distribution. Known from the Ganga-Brahmaputra river system in India, Bangladesh, Nepal, and Bhutan. Arunachal Pradesh: Abundantly found between mid-reaches and foot hill areas.

Economic importance. This fish has both food and ornamental value. However, it has more demand in ornamental market.

Conservational status: The species is widely spread in the Ganga-Brahmaputra river basin. This species is very abundant and a common cactched from all major rivers of Arunachal Pradesh. IUCN status: Least concern (LC).

***Opsarius tileo* (Hamilton 1822)**

Fig. 18.

FIG. 18 *Opsarius tileo*, RGUMF-320, 104 mm SL: lateral view.

Cyprinus tileo Hamilton, 1822: 276, 385; type locality: Kosi River, Uttar Pradesh, India.

Opsarius tileo: Kottelat, 2013: 132 (valid).

Material examined. RGUMF-320, 2 exs., 102–104 mm SL, Chatjo River in Tirap district, Arunachal Pradesh.

Vernacular names in Arunachal Pradesh. *Rebio Tapio* (in Nyishi), *Orpu* (in Galo), *Sepung* (in Adi).

Diagnosis. Snout pointed, mouth-cleft obliquely extended posteriorly underneath the middle of the orbit, a pair of rudimentary maxillary barbel present, three or more rows of vertically arranged spots or blotches on the flank, and lateral line scales 70–75.

Coloration. In live specimens, head and body dorsally bluish black and silvery on sides and belly, 3–4 rows of bluish spots or blotches on the flank, the uppermost row consists of about 13–15 blotches more prominent than the lower rows, interspaces more or less wide than the blotches. Pectoral, pelvic, and anal fin are yellowish. Caudal fin and dorsal fin grayish, except golden-yellow distal quarter of second simple to third dorsal branched rays. In 10% formalin, body dark brown dorsally, dull silvery on flank and belly, with gray or dark brown spots on the sides.

Distribution. Known from the Ganga-Brahmaputra river drainage in India, Bangladesh, and Nepal. Arunachal Pradesh: Nath and Dey (2000) reported from all major rivers of the state.

Economic importance. This fish has both food and ornamental value.

Conservational status: Earlier, this species was a very common in Arunachal Pradesh (Nath and Dey, 2000). This fish is now very rare in the state. We encountered this species only once from Chatjo River in Tirap district recently (January, 2017) during our sampling in major rivers of the state for the last 13 years (2004–17). IUCN status: Least concern (LC).

8. Genus: *Barilius* Hamilton, 1822

Barilius (subgenus of *Cyprinus*) Hamilton 1822: 266, 384. Type species: *Cyprinus barila* Hamilton 1822. Type by subsequent designation.

Body moderately elongated and compressed, shallow to deep; head pointed and often with tubercles; terminal mouth, obliquely directed upwards, symphyseal knob not well developed; gape of mouth not extending behind anterior rim of orbit; eyes large and superior; lip simple and thin; origin of dorsal fin located behind pelvic fin insertion; often extending over anal fin; muscular pads often present in front of bases of pectorals; 1–2 pairs of barbels; caudal fin forked.

Key to species of *Barilius*:

1a. Each scale on body dotted with blackish spots, each scales on lateral line dotted with a colon (:) like punctuation markings............................. *B. bendelisis*
1b. Scales not dotted with spots...2
2a. Predorsal scales 25 or more, lateral line with 60–70 scales .... *Barilus shacra*
2b. Predorsal scales 21–23, lateral line with 41–44 scales.........................3
3a. 14–17 blackish bars on flanks, head length equal to pectoral-fin length.............................. *Barilus barila*
3b. 10–14 blackish bars on flanks, head longer than pectoral fin.................. ....*B. vagra*

***Barilius bendelisis* (Hamilton 1807)**
Fig. 19.

FIG. 19 Lateral views of *Barilius bendelisis*: (A) female, RGUMF-13, 178 mm SL, and (B) male, RGUMF-14, 140 mm SL.

Cyprinus bendelisis Hamilton, 1807: 345, pl. 32; type locality: Vedawati stream, head waters of Krishna River near Heriuru, Mysore, India.

Barilius bendelisis: Vishwanath et al., 2007: 21 (valid).

Materials examined. RGUMF-13–14 & 16–17, 21 exs., 60.0–180 mm SL, Dikrong at Doimukh; Poma at Ramgath; Ranga at Kimin; Papum Pare District, Arunachal Pradesh.

Vernacular names in Arunachal Pradesh. *Rebio Tapio/Ngotabalum* (in Nyishi), *Orpu* (in Galo), *Sepung* (in Adi).

Diagnosis. Lateral line scales with two black spots at their bases and other scales on body with black spots, 8–12 bars on flank which touch the lateral line, 18–21 predorsal scales, and 40–45 lateral line scales.

Coloration. Body and head dorsally blackish, silvery on sides and belly, 8–12 vertical bars descending toward lateral line, indistinct in adults; lateral line scales with two black spots at their bases and other scales with single spot. Pectoral, pelvic, and anal fin yellowish, dorsal and caudal fins grayish.

Distribution. Widely distributed: known from India (both the northern and peninsular India), Bangladesh, Nepal, Bhutan, Myanmar, Pakistan, Thailand, and Sri Lanka. Arunachal Pradesh: abundantly found in all major rivers of the state.

Remarks. This species is sexually dimorphic, males possess tubercles on snout tip and sides, lower jaw, and on scales. Dark brown spots on scales are also more prominent in mature male than those of mature females. Vertical color bands on flanks are indistinct in males. The pectoral fin of male is provided with a characteristic muscular pad in front of its bases.

Economic importance. This fish has both food and ornamental value.

Conservational status: Widely spread species and a dominant species of all catches in all the streams and rivers of Arunachal Pradesh. IUCN status: Least concern (LC).

***Barilius vagra* (Hamilton 1822)**

Fig. 20.

FIG. 20 *Barilius vagra*, RGUMF-18, 183 mm SL: lateral view.

Cyprinus vagra Hamilton, 1822: 269, 385; type locality: Ganga River, Patna, India.

B. vagra: Vishwanath et al., 2007: 23 (valid).

Materials examined. RGUMF-18, 21 exs., 173–190 mm SL, Ranga at Kimin, Papum Pare District, Arunachal Pradesh (AP). RGUMF-unregistrated, 9 exs., 55.2–97.0 mm SL, Lohit River at Tezu, Lohit district, AP.

Vernacular names in Arunachal Pradesh. *Rebio Tapio/Ngotabalum* (in Nyishi), *Orpu* (in Galo), *Sepung* (in Adi).

Diagnosis. Body with 10–14 short vertical bars on flank, interbars thrice as wide as vertical bar, first, barbels two pairs: one each of rostral and maxillary, rostral barbel longer than maxillary, lateral line scales 41–42 + 2–3, and predorsal scales 20–22.

Coloration. In live ground body color silvery with 10–14 bluish-black bars on sides. In 10% formalin, over all the body is brownish with grayish bars on sides.

Distribution. The species is widely distributed in the Ganga-Brahmaputra River drainage of India, Nepal, and Bangladesh. Reports of this species from Afghanistan, Sri Lanka, and Pakistan need verification. In Arunachal Pradesh: abundantly found in the lower reaches of all major rivers of the state.

Remarks. This species is similar to *B. barila* in having two pairs of barbles, but can be easily distinguished from the later by its slender (vs. robust) body.

Economic importance. This fish has both food and ornamental value.

Conservational status: A widely spread species and a dominant species on all catches in various streams and rivers of Arunachal Pradesh. IUCN status: Least concern (LC).

***Barilius barila* (Hamilton, 1822)**

Fig. 21.

FIG. 21 *Barilius barila*, RGUMF-344, 56.9 mm SL: lateral view.

Cyprinus barila Hamilton, 1822: 267, 384; type locality: rivers of Northern Bengal, India.

B. barila: Kottelat, 2013: 81 (valid).

Materials examined. RGUMF-344, 1 ex., 56.9 mm SL, Tengapani River in Namsai, Namsai district, Arunachal Pradesh (AP). RGUMF-unregistered, 3 exs., 72.5–96.4 mm SL, Naharlagun local fish market, Papum Pare district (AP). RGUMF-unregistered, 2 exs., 53.6–56.8 mm SL, Chatjo River in Tirap district, Arunachal Pradesh.

Vernacular names in Arunachal Pradesh. *Rebio Tapio/Ngotabalum* (in Nyishi), *Orpu* (in Galo), *Sepung* (in Adi).

Diagnosis. Body with 14–17 vertical bars on flank extending from back to lateral line, head length equal to or slightly longer than the body depth at dorsal origin, barbels two pairs: one each of rostral and maxillary, pectoral fin as long as head, lateral line scales 42–44, and predorsal scales 21–23.

Coloration. In live, body with 14–15 bluish-black bars on sides. Body side silvery and dorsally yellowish brown. Pectoral fin and anal fin yellowish. Caudal-fin mid-rays yellowish. In preserved specimens (10% formalin), body brownish with 14–15 dark brown bars on flank.

Distribution. The species is widely distributed in Ganga-Brahmaputra river drainages of India, Nepal, and Bangladesh. In Arunachal Pradesh: abundantly found in the streams and rivers in Namsai and Tirap district of Arunachal Pradesh.

Remark. *B. barila* differs from *B. vagra* in having a robust (vs. slender) body and broader (vs. narrower) lateral bars.

Economic importance. This fish has both food and ornamental value.

Conservational status: The species is widely spread throughout the Ganga-Brahmaputra basin, however it is a very rare species in Arunachal Pradesh. IUCN status: Least concern (LC).

***Barilius shacra* (Hamilton 1822)**

Fig. 22.

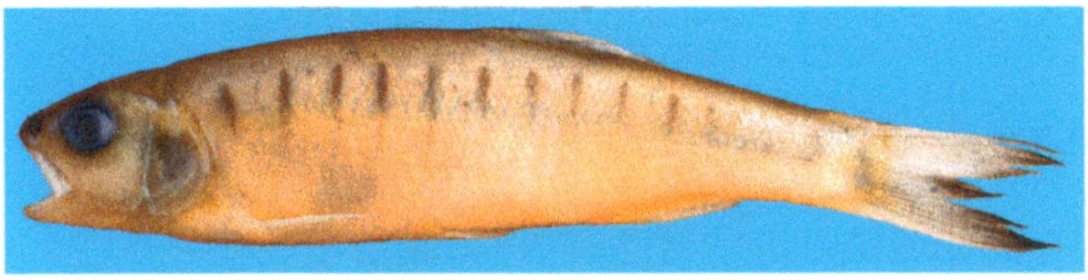

FIG. 22 *Barilius shacra*, RGUMF-434, 75.1 mm SL: lateral view.

Cyprinus shacra, Hamilton 1822: 271, type locality: Kosi River, Uttar Pradesh, India.

Barilius shacra: Mayanglambam & Vishwanath 2012: 2363 (valid).

Material examined. RGUMF-434, 04 exs., 70.5–76.9 mm SL, Dikrong river at Doimukh, Papum Pare district, Arunachal Pradesh.

Vernacular names in Arunachal Pradesh. *Rebio Tapio/Ngotabalum* (in Nyishi), *Orpu* (in Galo), *Sepung* (in Adi).

Diagnosis. Body with 10–12 incomplete bars along flank, scales minute, 25 predorsal scales, 60–70 lateral line scales, 2 pairs of barbels, axillary process distinct, caudal deeply forked.

Distribution. Known from the Ganga-Brahmaputra River drainage in India and Bangladesh. In Arunachal Pradesh: see material examined.

Coloration. In live, olive above and yellowish pink on sides and beneath, 10–12 incomplete blackish vertical bars descending from the back to lateral line. Dorsal fin with blackish bar.

Economic importance. This fish has both food and ornamental value.

Conservational status: A widely spread species and a dominant species in all catches in streams and rivers of Arunachal Pradesh. IUCN status: Least concern (LC).

9. Genus: *Laubuka* Bleeker, 1859

Laubuka Bleeker 1859: 261. Type species: *Cyprinus (Laubuka) laubuca* Hamilton 1822. Type by subsequent designation.

Body compressed, deep or moderate; mouth small directed obliquely or almost vertically upwards; cleft of the mouth reaching beneath to anterior rim of orbit, never extends to beneath or beyond the eye; eyes large; lacking symphysial process on lower jaw; barbels absent; lateral line complete or incomplete, or absent; 14 precaudal vertebrae, 31–37 + 1–2 lateral-line scales, 7–11 scales in transverse line on body and 14–20 branched anal fin rays.

***Laubuka laubuca* (Hamilton 1822)**

Fig. 23.

FIG. 23 *Laubuka laubuca*, RGUMF-362, 35.4 mm SL: lateral view.

Cyprinus laubuca Hamilton, 1822, 260, 384; type locality: Northern Bengal, India.

Laubuka laubuca: Kottelat, 2013: 119 (valid).

Material examined. RGUMF-362, 2 exs. 35.4–40.5 mm SL, Noadhing River, Arunachal Pradesh.

Vernacular names in Arunachal Pradesh. *Dori kana* (in Nyishi & Galo), *Garyong* (in Adi).

Diagnosis. A narrow indistinct grayish midlateral line on the body, more distinct on the posterior half of body, presence of a humeral and a caudal-peduncle spot; no tubercles on lower jaw; abdomen keeled only in between and behind pelvic fins; lateral line complete with 31–33 + 1–2 scales, transverse scales row from dorsal fin origin to pelvic-fin insertion with 6½/1/2½; fin counts: ii.8½ dorsal-fin rays, i.11–12 pectoral-fin rays, i.5–6 branched pelvic-fin rays, and iii.18½–20½ anal-fin rays.

Distribution. Known from India, Bangladesh, Nepal, and Pakistan. Records of this species from Myanmar are probably due to misidentification of *Laubuka khujairokensis*. In Arunachal Pradesh: abundantly found in foot hill areas.

Coloration. In live, body side silvery, dorsally darker, a narrow midlateral grayish line runs along the flank which becomes more prominent on posterior half of body, humeral and mid-peduncular spot blackish.

Remarks. This species differs from *Laubuka brahmaputraensis*, the only congener in the Brahmaputra basin, with its deeper (vs. slender) body, and having (vs. lacking) a blackish midlateral line and a caudal-peduncular spot.

Economic importance. Due to its small body size and attractive coloration, this fish has high ornamental value.

Conservational status: This fish is widely spread and abundantly occur in all foot hill streams and rivers of Arunachal Pradesh. IUCN status: Least concern (LC).

10. Genus: *Chela*, Hamilton, 1822

Chela (subgenus of Cyprinus) Hamilton 1822: 258, 383. Type species: Fem. *Cyprinus* (*Chela*) *cachius* Hamilton 1822. Type by subsequent designation.

Body compressed, deep or moderately; mouth small directed obliquely or almost vertically upwards; pelvic fin insertion close to pectoral base than to anal base, outer pelvic rays produced. Further the genus is diagnosed in having 17 precaudal vertebrae; lateral line scales 56–61 + 3–4; scales in transverse line of body with 17–19; 21–23 anal-fin rays.

***Chela cachius* (Hamilton 1822)**

Fig. 24.

FIG. 24 *Chela cachius*, RGUMF-168, 42.0 mm SL: lateral view.

Cyprinus cachius Hamilton, 1822: 258, 384; type locality: River Ganges above the commencement of the Delta, India.

Chela cachius: Kottelat, 2013: 87 (valid).

Material examined. RGUMF-168, 14 exs., 35.0–45.0 mm SL, Poma at Ramgath, Papum Pare district, Arunachal Pradesh.

Vernacular names in Arunachal Pradesh. *Dori kana* (in Nyishi & Galo), *Garyong* (in Adi).

Diagnosis. Mouth terminal, outer rays of pelvic fin extends up to anal origin, no spot behind gill opening, lower jaw prominent, barbels absent, lateral line complete with 56–61 + 3–4 scales, scales in transverse line on body with 17–19, predorsal scales 28–33, circumpenduncular scales 16–22; 21–23 branched anal fin rays, and 3–4 branched pelvic-fin rays.

Coloration. In live, over all the body silvery, darker above due to heavy concentration of melaniphores pigments. In 10% formalin, overall body color brownish.

Distribution. Widely distributed: India, Bangladesh, Nepal, and Pakistan. Occurrence of this species in Myanmar is unlikely. In Arunachal Pradesh: abundantly found in the foot hill streams and rivers of the state.

Remarks. This species is often used as bait for angling purpose.

Economic importance. Due to its small size and beautiful coloration, this fish has high ornamental value.

Conservational status: This fish is widely spread and abundantly occurring in all foot hill streams and rivers of Arunachal Pradesh. IUCN status: Least concern (LC)

11. Genus: *Danio* Hamilton, 1822

Danio (subgenus of Cyprinus) Hamilton 1822: 321, 390. Type species: Cyprinus (*Danio*) *dangila* Hamilton 1822. Type by subsequent designation.

Body compressed; abdomen rounded; mouth anterior and not protractile, mouth cleft shallow and obliquely directed upwards, lower jaw bears a symphyseal knob; barbels two pairs; horizontal stripes on sides sometimes with ring patterns, two or more pigment stripes on the caudal-fin rays; dorsal fin inserted opposite to interspaces between anal and pelvic fins, its origin close to caudal fin base than to tip of snout.

Key to species of *Danio*:

1a. Ring shaped markings along flanks, ring pattern not extending onto caudal peduncle and lateral line complete with scale count 32–36..*D. assamila*

1b. No ring shaped markings along flanks, lateral line absent .... *D. rerio*

***Danio assamila* Kullander 2015**
Fig. 25.

FIG. 25 *Danio assamila*, RGUMF-322, 56.7 mm SL: lateral view.

D. assamila Kullander, 2015: 362, Fig. 1c–d; type locality: Assam, Brahmaputra River drainage, tributary of Dibru River, 2 km north of Digholtrang, Amguri, India.

Danio dangila: Nath & Dey, 2000: 21, Fig. 14 (description).

D. assamila: Phurin et al., 2017: 426 (valid).

Materials examined. RGUMF-322, 3 exs., 40.8–56.7 mm SL, Chatjo River (R.) at Deomali, a tributary of Burhi-Dihing R., Brahmaputra basin (Bb.), Tirap district, Arunachal Pradesh (AP); collected by Phurin Songtheng, November, 2016. RGUMF-21, 9 exs., 42.0–57.0 mm SL, Poma R. in Papum Pare district, Bb., AP. RGUMF-132, 15 exs., 44.3–73.4 mm SL, Singen R. (upper reaches of Simen R., Bb.), East Siang District, AP.

Vernacular names in Arunachal Pradesh. *Tapio* (in Nyishi), *Tapo* (in Galo), *Garyong* (in Adi).

Diagnosis. First ray of pectoral and pelvic fins produced distally from rest of the rays, cleithral spot large and rounded or slightly oval, anterior interstripe Ia usually present between the two series of horizontally elongated ring shaped markings along flank, ring chain pattern not extending onto caudal peduncle, and lateral line complete with scale count 32–36.

Coloration. In live, body silvery, with cleithral dark blue spot and bordered entirely or partially by a lighter zone. In 10% formalin, ground color is dull yellowish, lighter abdominally or dull whitish along ventral midline. Cleithral spot blackish, rounded, equal to pupil size, slightly ovate in some specimens, narrower ventrally. Mid-dorsal stripe brownish. Lateral P, P+1, P+2, P+3, P−1, and P−2 stripes dark brown; P−2 stripe diffuse brownish in some specimens.

Distribution. Known only from the Brahmaputra drainage in Northeast India: Assam and Arunachal Pradesh.

Remark. This species is referred to the Chain danio group because of its chain-like marking pattern on body sides (Kullander, 2015). *D. dangila* differs from *D. assamila* in having vertically extended (vs. rounded) cleithral spot, rounded (vs. horizontally elongated) markings on flank, and ring chain pattern usually extending onto caudal peduncle (not extending to caudal peduncle). Earlier reports of *D. dangila* from Arunachal Pradesh are probably due to misidentification of *D. assamila*.

Economic importance. This fish has high ornamental value due to its small size and distinctive coloration.

Conservational status: A species of the upper Brahmaputra river basin, commonly inhibits in all the foothill streams and rivers of Arunachal Pradesh and Assam, state of Northeast India. IUCN status: Not assessed (NA).

***Danio rerio* (Hamilton 1822)**

Fig. 26.

FIG. 26 *Danio rerio*, RGUMF-22, 25.3 mm SL: lateral view.

Cyprinus rerio Hamilton, 1822: 323, 390; type locality: Kosi river, India.

Danio rerio: Fang, 2003: 717 (valid).

Material examined. RGUMF-22, 16 exs., 10.0–34.0 mm SL, Dikrong River at Doimukh, Papum Pare district, Arunachal Pradesh.

Vernacular names in Arunachal Pradesh. *Tapio* (in Nyishi), *Tapo* (in Galo), *Garyong* (in Adi).

Diagnosis. Lateral line absent, barbels two pairs: one pair each of rostral and maxillary, rostral short and maxillary extend up to operculum; I+2, I+1, I, and I−1 bluish horizontal stripes are continuous and marginally smooth, and anal fin with ii,11–12 rays.

Coloration. In 10% formaldehyde, body background color whitish with a mid-dorsal and 4–5 dark brown lateral stripes. I (middle) stripe uniformly wide, I+1 and I−1 slightly narrower than I stripe. I+2 stripe extend from dorsolateral side of head to the level of dorsal fin base end, I – 2 stripe prominent between the origin of pelvic to middle of anal-fin base. Lateral stripes: I, I + 1, and I – 1 continuous to caudal fin. Anal fin with 3–4 bands across their rays.

Distribution. India, Nepal, Bangladesh, and Bhutan. In Arunachal Pradesh: abundantly found in all foot hill areas.

Remarks. This species is the most studied fish among vertebrates. Its genome has been fully sequenced.

Economic importance. *D. rerio* has high ornamental value because of its small size and beautiful coloration. It is also popularly known as zebra *danio*.

Conservational status: This fish is widely spread and abundantly occur in all foot hill streams and rivers of Arunachal Pradesh. IUCN status: Least concern (LC).

12. Genus: *Devario* Heckel, 1843

Devario Heckel 1843: 1015. Type species: *Cyprinus devario* Hamilton 1822. Type by absolute tautonymy.

Dorsal body profile slightly straighter than ventral profile; head and body laterally compressed; abdomen rounded; mouth supra-terminal, cleft of mouth shallow, not protractile; a rounded symphyseal knob present on lower jaw, with a corresponding shallow groove on inner margin of upper jaw; body depth greatest at pelvic-fin origin; flanks with horizontal or vertical color markings; two pairs of barbels: maxillary barbel very short, rostral barbel not reaching to anterior rim of orbit.

Key to the species of *Devario*:

1a. Anterior rim of orbit without a backward curve spine, anterior part of body with scattered bars enclosing yellowish to golden patches.................... ....*D. devario*

1b. Anterior rim of orbit with a backward curve spine, anterior part of body lacking scattered bars...2

2a. Circumpeduncular scales 10–12, anal-fin branched rays 14–15 .... *D. aequipinnatus*

2b. Circumpeduncular scales 14 and anal-fin branched rays 16–17 .... *D. assamensis*

***Devario aequipinnatus* (McClelland 1839)**
Fig. 27.

FIG. 27 *Devario aequipinnatus*, RGUMF-26, 65.0 mm SL: lateral view.

Perilampus aequipinnatus M'Clelland, 1839: 393, pl. 60, Fig. 1; type locality: India: Assam.

Devario aequipinnatus: Kottelat, 2013: 98 (valid).

Material examined. RGUMF-23–26, 20 exs., 32.0–75.0mm SL, Kele at Poma, Labia in Papum Pare district, Arunachal Pradesh.

Vernacular names in Arunachal Pradesh. *Tapia/Tapio* (in Nyishi), *Tapo* (in Galo), *Garyong* (in Adi).

Diagnosis. Body and head compressed, belly rounded. A backward curve preorbital spine originated from lachrymal bone, 35–37 lateral line scales, and 14–15 predorsal scales present. Four bluish stripes P+2, P+1, P, and P−1 (in sequence from uppermost to bottom) are separated by yellowish interstripes I+2, I+1, I−1, I−2. P and P−1 stripes are broader than other stripes, P stripe slightly larger than P−1, and anal fin with ii–iii,14–5 rays.

Coloration. See diagnosis.

Distribution. India, Nepal, Bangladesh, Bhutan, and Sri Lanka. In Arunachal Pradesh: abundantly found in foot hill areas.

Remark. *D. aequipinnatus* differs from its closest congener *D. assamensis* by its fewer circumpeduncular scales (10–12 vs. 14) and anal-fin branched rays (14–15 vs. 16–17). The two species easily differentiates from each other by their body color pattern (see above).

Economic importance. A hardy aquarium fish because of its small size and beautiful coloration.

Conservational status: This species is abundantly encountered in all the low-land water bodies of the Ganga-Brahmaputra basin. IUCN status: Least concern (LC).

***Devario devario* (Hamilton 1822)**

Fig. 28.

FIG. 28 *Devario devario*, RGUMF-27, 48.0mm SL: lateral view.

Cyprinus devario Hamilton, 1822: 341, 393, pl. 6; Fig. 94, type locality: Bengal, India. *Devario devario*: Vishwanath et al., 2007: 32 (valid).

Material examined. RGUMF-27, 5 exs., 48.0–60.0mm SL, Ranga at Kimin, Papum Pare district, Arunachal Pradesh.

Vernacular names in Arunachal Pradesh. *Tapia/Tapio* (in Nyishi), Tapo (In Galo), *Garyong* (in Adi).

Diagnosis. Infraorbital spine absent, lateral line complete with 45–52 scales, dorsal fin with ii,15–17 rays, and anal fin with 16–19 branched rays; P+2, P+1, P, P−1 horizontal stripes separated in sequence by I+2, I+1, I−1 on posterior half of the body, anteriorly with scattered bars enclosing yellowish to golden patches, P stripe broad and continuous to the end of caudal fin, and anal fin with ii–iii,16–19 rays.

Coloration. See diagnostic character above.

Distribution. The species is known from India, Nepal, and Bangladesh. In Arunachal Pradesh: abundantly found in foot hill areas.

Remarks. Absence or presence of barbel for this species have no taxonomic importance because its maxillary barbel is rudimentary and may often be absent entirely.

Economic importance. A fish of high ornamental value because of its small size and beautiful coloration.

Conservational status: This species is abundantly encountered in all the low-land water bodies of the Ganga-Brahmaputra basin. IUCN status: Least concern (LC).

***Devario assamensis* (Barman 1984)**

Fig. 29.

FIG. 29 *Devario assamensis*, RGUMF-361, 58.6 mm SL: lateral view.

Danio assamensis Barman 1984: 163; type locality: Streamlets near Tangla, Darrang district, Assam, India.

Devario assamensis: Vishwanath et al., 2007: 32 (description, valid).

Material examined. RGUMF-361, 01 ex., 58.6 mm SL, Tengapani River in Namsai, Namsai district, Arunachal Pradesh.

Vernacular names in Arunachal Pradesh. *Tapia/Tapio* (in Nyishi), *Tapo* (in Galo), *Garyong* (in Adi).

Diagnosis. Two pairs of barbels present**,** anterior rim of orbit with a small backwardly projecting spine, lateral line complete with 36–37 scales, predorsal scales 15–16, circumpendecular scales 14; scales on transverse row between the dorsal fin and the lateral line 7½, row between the pelvic-fin origin and the lateral line 2½; dorsal fin with ii,12 rays, anal fin with ii,16–17 rays. P stripes are separated by two distinct narrow pale interspaces; a black cleithral spot present behind the upper edge of gill opening.

Coloration. In live, flank with pale bluish P+1, P and P−1 stripes separated in sequence by I+1 and I−1 interstripes. P stripe width is twice that of P+1 or P−1.

Distribution. Known only from India: Assam and Arunachal Pradesh.

Remarks. *D. assamensis* differs from *D. devario* in having, fewer lateral line (36–37 vs. 45–52) scales, dorsal fin branched rays 12 (vs. 15–17), and lacking (vs. having) P+2 stripe.

Economic importance. *D. assamensis* has high ornamental value because of its small size and beautiful coloration.

Conservational status: This species is restricted in the Brahmaputra basin only. IUCN status: Vulnerable (Vu).

13. Genus: *Esomus* Swainson, 1839

Esomus (subgenus of *Leuciscus*) Swainson 1839: 185, 285. Type species: *Esomus vittatus* Swainson 1839 (=*Cyprinus danrica* Hamilton 1822). Type by monotypy.

Body elongated and compressed; head small; abdomen rounded; mouth small, obliquely directed upwards; lips thin, lower jaw prominent, symphyseal knob absent; barbels two pairs, maxillary pair extended up to anal fin; dorsal fin inserted opposite to the interspace between the anal and pelvic fins; caudal fin forked; scales moderate in size, lateral line may be absent or present.

***Esomus danrica* (Hamilton 1822)**

Fig. 30.

FIG. 30 *Esomus danrica*, RGUMF-20, 39.0mm SL: lateral view.

Cyprinus danrica Hamilton, 1822: 325, 390; type locality: ponds and ditches of Bengal, India.

Esomus danrica: Kottelat, 2013: 102 (valid).

Material examined. RGUMF-20, 3 exs., 37.0–41.0mm SL, Dikrong River at Doimukh, Papum Pare district, Arunachal Pradesh.

Vernacular names in Arunachal Pradesh. *Doricana* (in Nyishi & Galo), *Garyong* (in Adi).

Diagnosis. Body slender and compressed with a broad midlateral blackish band extending from behind the orbit to the base of caudal fin. Head pointed and mouth upturned. Barbels two pairs, one each of maxillary and rostral, maxillary barbels extend up to anal fin. Dorsal fin inserted opposite to the interspace between the anal and pelvic fins, bearing ii,6 rays. Lateral line incomplete with 3–6 pierced scales, predorsal scales 16–17, and circumpeduncular scales 14.

Coloration. In live, body silvery white, dorsally darker. A dark greenish midlateral stripe of two-scale width extending from slightly behind orbit to base of caudal fin. A golden yellowish line runs immediately above the midlateral stripe. Pelvic fin reddish and other fins hyaline.

Distribution. The species is known from India, Nepal, Bangladesh, and Myanmar. In Arunachal Pradesh: abundantly found in all the water bodies of foot hill areas.

Remark. It is also a delicacy dish in Northeast India.

Economic importance. It has high ornamental value because of its small size and beautiful coloration.

Conservational status: A widely spread species known from both the Ganga-Brahmaputra and Chindwin-Irrawaddy river basins. This fish is commonly found in small streams, pound, paddy field. IUCN status: Least concern (LC).

14. Genus: *Raiamas* Jordan, 1919

Raiamas Jordan 1919: 344. Type species: *Cyprinus bola* Hamilton 1822. Type by being a replacement name.

Body elongated and compressed; head pointed; mouth terminal and wide, obliquely directed upwards; maxilla extends beyond posterior orbital rim; lower jaw with a symphyseal knob fitting into a median nodes of upper jaw; barbels absent or with a rudimentary pair; dorsal fin inserted behind pelvic-fin origin.

***Raiamas bola* (Hamilton 1822)**

Fig. 31.

FIG. 31 *Raiamas bola*, RGUMF-19, 130 mm SL: lateral view.

Cyprinus bola Hamilton, 1822: 274, 385; type locality: Brahmaputra River, India.

Raiamas bola: Vishwanath et al., 2007: 24 (description, valid).

Material examined. RGUMF-19, 4 exs., 68.0–130 mm SL, Poma river at Ramgath, Papum Pare District, India.

Vernacular names in Arunachal Pradesh. *Ngota Bolam* (in Nyishi), *Bola* (in Adi & Galo).

Diagnosis. Mouth obliquely upturned and cleft extending beyond vertical through posterior margin of orbit. Lateral line scales 85–95. Numerous oblong diffuse spots in 2–4 rows on flank.

Coloration. In live, body silvery with bluish spots in 2–4 rows on sides. In juvenile, the uppermost row of bluish spots is more prominent than the lower and more or less straight.

Distribution. Known from India, Nepal, Bangladesh and Bhutan. In Arunachal Pradesh: abundantly found up to mid-altitude in all major rivers of the state.

Remark. This species is regarded as carnivorous fish.

Economic importance. It has both food and ornamental value.

Conservational status: A widely spread species occurring throughout the Ganga-Brahmaputra river basin. IUCN status: Least concern (LC).

15. Genus: *Rasbora* Bleeker, 1859

Rasbora Bleeker 1859: 361, 371. Type species: *Leuciscus cephalotaenia* Bleeker 1852.

Body elongated and moderately compressed; snout pointed; mouth terminal, directed obliquely upward; lower jaw projected with a symphyseal knob; barbel absent, lateral line complete.

Key to the species of *Rasbora*:

1a. Caudal fin with a well-defined blackish distal border, body midlateral stripe very thin and indistinct anteriorly.. *R. rasbora*
1b. Caudal fin lacking a blackish distal border, body midlateral stripe broad, and entirely distinct in length from snout to the base of caudal fin .... *R. daniconius*

***Rasbora daniconius* (Hamilton 1822)**
Fig. 32.

FIG. 32 *Rasbora daniconius*, RGUMF-169, 46.0mm SL: lateral view.

Cyprinus daniconius Hamilton, 1822: 327, 391, pl. 15, Fig. 89; type locality: southern Bengal, India.

Rasbora daniconius: Kottelat, 2013: 152 (valid).

Material examined. RGUMF-169, 3 exs., 41.0–48.0mm SL, Ranga River at Kimin, Papum Pare district, Arunachal Pradesh.

Vernacular names in Arunachal Pradesh. *Doricana* (in Nyishi & Galo), *Garyong* (in Adi).

Diagnosis. A black broad midlateral stripe extending from preorbital region to the base of caudal fin. Lateral line complete and concave with 30–34 scales. Barbel absent. Predorsal scales 14.

Coloration. In live, olive on back and silvery on sides with a blue black broad midlateral stripe running from the preorbital region to the base of caudal fin.

Distribution. Widely distributed: India, Nepal, Bangladesh, Bhutan, Sri Lanka, Pakistan, Laos, Thailand, Lao, Indonesia, Thailand, Vietnam, Malaysia, and Myanmar In Arunachal Pradesh: abundantly found up to mid-altitude.

Remark. This species is commonly known as slender-rasbora or blackline-rasbora.

Economic importance. It has both food and ornamental value.

Conservational status: A widely spread common fish species occurring in multiple river basins in the southeast Asia. IUCN status: Least concern (LC).

***Rasbora rasbora* (Hamilton 1822)**

Fig. 33.

FIG. 33 *Rasbora rasbora*, RGUMF-190, 56.4 mm SL: lateral view.

Cyprinus rasbora, Hamilton, 1822: 329, 391, pl. 2, Fig. 90; type locality: Bengal, India.

Rasbora rasbora: Vishwanath et al., 2007: 37 (description, valid).

Material examined. RGUMF-190, 1 ex., 56.4 mm SL, Dikrong River, Papum Pare District, Arunachal Pradesh.

Vernacular names in Arunachal Pradesh. *Doricana* (in Nyishi & Galo), *Garyong* (in Adi).

Diagnosis. A blackish longitudinal midlateral stripe, barbel absent, lateral line complete with 28–31 scales, caudal fin with well-defined blackish distal border, and predorsal scales 12.

Coloration. In live, body silvery with a faint streak along the side from the eyes to the caudal fin base. Distal portion of caudal fin lobes with broad blackish marking interconnected by two blackish distal margins of inner rays.

Distribution. Widely distributed: India, Bangladesh, and Pakistan. In Arunachal Pradesh: abundantly found in foot hill areas.

Remark. This species differs from *R. daniconius* in having a narrow faint (vs. broad and distinct) midlateral stripe, and fewer predorsal scales (12 vs. 14).

Economic importance. It has both food and ornamental value.

Conservational status: A widely spread common fish species occurring in multiple river (Ganga-Brahmaputra and Indus River) basins. IUCN status: Least concern (LC).

16. Genus: *Salmostoma* Swainson, 1839

Salmostoma Swainson 1839: 184. Type species: *Cyprinus oblongus* Swainson 1839 (= *Cyprinus bacaila* Hamilton 1822). Type by subsequent designation

Body elongated and compressed; abdomen keeled; mouth oblique, lower jaw longer with a well-developed symphysial process and a notch in upper margin of corresponding jaw; broad suborbital bones; dorsal fin short, inserted opposite to anal or slightly ahead of anal origin; outer rays of pelvic fin elongated; lateral line complete.

Key to the species of *Salmostoma*:

1a. Mouth cleft go beyond anterior rim of orbit, 118–122 lateral line scales, 69–70 predorsal scales.. ....*S. bacaila*

1b. Mouth cleft not exceeds to anterior rim of orbit, 89–98 lateral line scales, 52–57 predorsal scales.. *S. phulo*

***Salmostoma bacaila* (Hamilton 1822)**
Fig. 34.

FIG. 34 *Salmostoma bacaila*, RGUMF-08, 70.3 mm SL: lateral view.

Cyprinus bacaila Hamilton, 1822: 265, 384, pl. 8, Fig. 76; type locality: Rivers of India.

Salmophasia bacaila: Vishwanath et al., 2007: 15 (description, valid).

Material examined. RGUMF-08, 5 exs., 46.0–78.0 mm SL, Dikrong at Doimukh, Papum Pare district, Arunachal Pradesh.

Vernacular names in Arunachal Pradesh. *Doricana* (in Nyishi & Galo), *Garyong* (in Adi).

Diagnosis. Mouth oblique, mouth cleft go beyond anterior rim of orbit, lateral line complete with 118–122, 19/6 transverse scale rows, anal-fin base length short with 10–14 branched rays, 17–21 rakers on first gill arch, and 69–70 predorsal scales.

Coloration. In live, body dorsally gray greenish, and side silvery with a shiny white broad midlateral band. All fins hyaline.

Distribution. India, Nepal, Bangladesh, Pakistan, and Afghanistan. In Arunachal Pradesh: abundantly found up to mid-altitude.

Remark. It is a surface feeder and regarded as a carnivorous fish (Talwar and Jhingran, 1991).

Economic importance. This species is preferred by people as a food fish. The species has high consumer preference for its food value.

Conservational status: A widely spread species occurring throughout the Ganga-Brahmaputra river basin. Least concern (LC)

***Salmostoma phulo* (Hamilton 1822)**

Fig. 35.

FIG. 35 *Salmostoma phulo*, RGUMF-328, 74.4 mm SL: lateral view.

Cyprinus phulo Hamilton, 1822: 263, 384; type locality: Ganges in Gangetic provinces.

Salmophasia phulo: Vishwanath et al., 2007: 15 (description, valid).

Materials examined. RGUMF-328, 03 exs., 47.0–74.4 mm SL, Poma River, Papum Pare district, Arunachal Pradesh. RGUMF-unregistered, 05 exs., 52.1–65.1 mm SL, Sinki river, Lower Dibang valley district, Arunachal Pradesh.

Vernacular names in Arunachal Pradesh. *Doricana* (in Nyishi & Galo), *Garyong* (in Adi).

Diagnosis. Mouth oblique, mouth cleft does not exceed to anterior rim of orbit, lower jaw longer with a distinct symphysial knob and upper jaw with a corresponding depression, lateral line complete with 89–98 scales, anal-fin base long with 17–19 branched rays, 15–16/5–6 transverse scale rows, 19 gill rakers on first arch, and 52–57 predorsal scales.

Coloration. In live, body silvery with a bright shiny white lateral band.

Distribution. Known from India, Nepal, and Bangladesh. In Arunachal Pradesh: Dikrong, Ranga, Senkhi Rivers.

Remarks. This species is considered as a food fish.

Conservational status: A widely spread species occurring throughout the Ganga-Brahmaputra river basin. Not assessed (NA)

17. Genus: *Securicula* Günther, 1868

Securicula (subgenus of *Chela*) Günther 1868: 332. Type species: *Cyprinus gora* Hamilton 1822. Type by subsequent designation.

Body strongly compressed and fairly elongate; eyes large; mouth slightly upward, lower jaw longer, a prominent cartilaginous knob on the lower jaw; abdominal edge sharply keeled from below the pectoral fin to the anal fin;

abdominal keel anteriorly supported by an extension of pectoral girdle; a muscular mass covered by skin and scutes extends on to the dorsal face of head to nostril; pelvic fins with well-developed axillary scales; lateral line complete with about 140 scales; caudal fin forked, lower lobe longer than the upper lobe.

Securicula gora **(Hamilton 1822)**

Fig. 36.

FIG. 36 *Securicula gora*, RGUMF-327, 117mm SL: lateral view.

Cyprinus gora, Hamilton 1822: 263, 384; type locality: Brahmaputra River near Goyalpara [Goalpara?], India.

Securicula gora: Menon, 1999: 175: 329 (valid).

Material examined. RGUMF-327, 01 ex., 117mm SL, purchase from Naharlagun Bazar (fisherman's information: catches from Dikrong River), Papum Pare district, Arunachal Pradesh.

Vernacular names in Arunachal Pradesh. Not available (Nyishi, Adi, & Galo).

Diagnosis. Body elongated and compressed, mouth oblique scales very small, body scales covering extended forward on the head up to above the nostrils, abdomen lacking scales and sharply keeled from the origin of anal fin to below opercle, lateral line scales 120–160, dorsal fin inserted in advance of anal, and anal fin with 13–15 branched rays.

Coloration. In live, body uniformly bright silvery. In 10% formalin, brownish on head and dorsum, silvery below.

Distribution. Widely distributed in India, Bangladesh, Nepal, and Pakistan. In Arunachal Pradesh: abundantly found in foothill streams and rivers.

Remarks. This species is a predatory surface feeder and feeds on insects, larvae, and crustaceans.

Economic importance. This species is considered as a food fish.

Conservational status: A widely spread species occurring throughout the Ganga-Brahmaputra river basin. Least concern (LC)

Subfamily: Labeoninae

18. Genus: *Bangana* Hamilton, 1822

Bangana (subgenus of *Cyprinus*) Hamilton 1822: 277, 385. Type species: *Cyprinus dero* Hamilton 1822. Type by subsequent designation.

Body subcylindrical; covered with cycloid scales; upper lip smooth and entire, separated from snout by a deep groove; upper jaw fully enclosed in upper lip, median part of upper lip covered slightly by rostral fold; lower jaw heavily cornified; post-labial groove widely uninterrupted and present only on sides of jaw; lower lip thin, a deep transverse groove separate the lower lip from mental region; dorsal fin with 10–12 branched rays, spinous ray absent.

***Bangana dero* (Hamilton 1822)**

Fig. 37.

FIG. 37 *Bangana dero*, RGUMF-56, 280.5 mm SL: (A) lateral view, (B) dorsal view of snout, and (C) ventral view of mouth part.

Cyprinus dero Hamilton, 1822: 277, 331, 585, pl. 22, Fig. 78; type locality: River Brahmaputra, India.

Bangana dero: Vishwanath et al., 2007: 76 (description, valid).

Materials examined. RGUMF-55–56, 5 exs., 105–226 mm SL, Dikrong river at Sagalee, Ranga river at Kimin in Papum Pare district, Arunachal Pradesh (AP). RGUMF-456, 280.5 mm SL, upper reaches of Subansiri River at Basar village, West Siang district, AP.

Vernacular names in Arunachal Pradesh. *Ngopii* (in Adi). *Ngopii yalik in Nyishi.*

Diagnosis. Snout conical with a deep groove and larges spiny tubercles, post-labial fold interrupted, one pair of short maxillary barbels present, upper lip partly covered by rostral fold, upper surface of lower lip studded with papillae; ii,9–10½ dorsal-fin rays; lateral line scales 41–43, lateral transverse scales 7/7, and circumpedencular scales 24.

Coloration. Body brownish with dull silvery belly. A prominent black spot present at the shoulder, just above the fifth lateral line scale.

Distribution. India, Bangladesh, Nepal and China. In Arunachal Pradesh: abundantly found up to mid-altitude.

Remarks. Some authors treated this species as a member of *Labeo*.

Economic importance. This species is a high priced food fish.

Conservational status: A widely spread species occurring throughout the Ganga-Brahmaputra river basin. Least concern (LC)

19. Genus: *Gibelion* Heckel, 1843

Gibelion Heckel 1843: 1014. Type species: *Cyprinus catla* Hamilton 1822. Type by subsequent designation.

Body moderately short, deep, and compressed; body deeper than head length; head broad and large; protruding lower jaw; mouth upturned, wide, terminal; upper lip absent, lower lip thick; snout bluntly rounded; grayish scales on its dorsal broad side and whitish on its belly; barbels absent, dorsal fin inserted ahead of pelvis fins, anal fin short, caudal fin forked, lateral line complete with 40–43 scales.

***Gibelion catla* (Hamilton 1822)**

Fig. 38.

FIG. 38 *Gibelion catla*, RGUMF-324, 240 mm SL: lateral view.

Cyprinus catla Hamilton, 1822: 287, 387, pl. 13 Fig. 81; type locality: Bengal and part of Bihar, India.

Gibelion catla: Kottelat, 2013: 107 (valid).

Material examined. RGUMF-324, 1 ex., 240 mm SL, purchase from Doimukh market, Papum Pare district, Arunachal Pradesh.

Vernacular names in Arunachal Pradesh. *Catla/Orpey* (in Nyishi, Adi, & Galo).

Diagnosis. Body: short and deep, head: broad and large, snout bluntly rounded, lateral line complete with 42–43 scales, upper lip absent, and lower lip very thick and continuous with a free posterior margin, anal fin small with 4–5 branched rays and barbels absent.

Coloration. In live, body silvery on flank and belly, dorsally brownish gray.

Distribution. India, Nepal, Bhutan, Bangladesh, and Pakistan. In Arunachal Pradesh: abundantly found in foot hill stream and rivers, usually culture in ponds.

Remarks. The species is one the indigenous fishes belonging to Indian Major Carps (IMC). This is the only species known under the genus *Gibelion*.

Economic importance. This fish is the most popular food fish in Asia.

Conservational status: A widely spread species occurring throughout the Ganga-Brahmaputra river basin. Least concern (LC)

20. Genus: *Cirrhinus* Oken, 1817

Cirrhinus Oken (ex Cuvier) 1817: 1182a, 1183. Type species: *Cyprinus cirrhosus* Bloch 1795. Type by monotypy.

Body elongated and compressed; dorsal profile more convex than abdomen profile; snout obtusely rounded with pores; upper lip discontinuous with the lower lip; lower jaw sharp and usually lacks cartilaginous covering inside; eyes moderately large, situated close to tip of snout than the posteriormost edge of operculum; barbel one or two pairs or absent; caudal peduncle short; dorsal fin inserted slightly anterior to pelvic fin, 2–4 nonosseous simple rays, followed by up to 15 branched rays; lateral line complete with 35–45 scales, caudal fin forked.

Key to the species of *Cirrhinus*:

1a. Flanks with 6–7 bluish horizontal bands, lateral line scales 34–38 .... *C. reba*

1b. Flanks lacking bands, lateral line scales 40–45...................... *C. mrigala*

***Cirrhinus mrigala* (Hamilton 1822)**
Fig. 39.

FIG. 39 *Cirrhinus mrigala*, RGUMF-59, 150 mm SL: lateral view.

Cyprinus mrigala Hamilton, 1822: 297, 389, pl. 6 Fig. 79; type locality: Ponds and freshwater rivers of Gangetic provinces, India.

Cirrhinus mrigala: Jayaram, 1999: 130–131 (description, valid).

Materials examined. RGUMF-59, 3 exs., 110–150 mm SL, Ranga River at Kimin, Papum Pare district, Arunachal Pradesh. RGUMF-unregistered, 01 ex., 150 mm SL, purchase from Doimukh market, Papum Pare district, Arunachal Pradesh.

Vernacular names in Arunachal Pradesh. *Mrigal/Orpey* (in Nyishi, Adi, & Galo).

Diagnosis. Mouth sub-inferior, upper lip entire and lower lip indistinct. Dorsal fin inserted ahead of pelvic-fin origin, with 3 simples and 12–13 branched rays. Scales rounded. Transverse scale rows above lateral line with 7–8, lateral line scales 40–45.

Coloration. In live, bright silvery white, dorsally brownish gray.

Distribution. India, Nepal, Bhutan, Bangladesh, and Pakistan. In Arunachal Pradesh: abundantly found in the foot hill streams and rivers, usually culture in ponds.

Remarks. The species is one of the indigenous fishes that belong to IMC. *C. mrigala* differs from *C. reba* by its greater number (40–45 vs. 34–38) of lateral line scales and scales rounded (vs. hexagonal).

Economic importance. This species is a popular food fish.

Conservational status: A widely spread species occurring throughout the Ganga-Brahmaputra river basin. Least concern (LC)

***Cirrhinus reba* (Hamilton 1822)**

Fig. 40.

FIG. 40 *Cirrhinus reba*, RGUMF-60, 145 mm SL: lateral view.

Cyprinus reba Hamilton, 1822: 283, 386; type locality: Rivers and ponds of West Bengal, India.

Cirrhinus reba: Shestha 2008: 60 (valid).

Material examined. RGUMF-60, 1 ex., 145 mm SL, Dikrong River at Doimukh, Papum Pare district, Arunachal Pradesh.

Vernacular names in Arunachal Pradesh. *Reba/Orpey* (in Nyishi, Adi, & Galo).

Diagnosis. Body: elongate, depth greater than head length. Mouth broad, upper lip entire, lower jaw with a thin cartilaginous cover. A pair of short rostral barbels present. Scale size moderate and hexagonal in shape, lateral line scales 34–38, transverse scale rows 7/5–6. Fin rays counts: dorsal ii–iii.8, anal iii.5, pectoral i.15, and pelvic i.8.

Coloration. In live, dark gray dorsally and silvery below, flank with 6–7 bluish horizontal bands.

Distribution. India, Pakistan, Nepal, Sri Lanka, and Bangladesh. In Arunachal Pradesh: abundantly found in the foot hills streams and rivers and also cultured in ponds.

Economic importance. This species is a popular food fish.

Conservational status: A widely spread species occurring throughout the Ganga-Brahmaputra river basin. IUCN status: Least concern (LC).

21. Genus: *Tariqilabeo* Mirza & Saboohi, 1990

Tariqilabeo (subgenus of *Labeo*) Mirza & Saboohi 1990: 405. Type species: Masc. *Labeo macmahoni* Zugmayer 1912. Type by original designation (also monotypic).

Body elongated and cylindrical; snout prominent and smooth; mouth inferior, upper lip prominent, continued in fold at corners of mouth; lower lip fleshy and fits into a shallow depression on horny sheath of lower jaw; margin of lower lip with lobate papillae.

***Tariqilabeo latius* (Hamilton 1822)**

Fig. 41.

FIG. 41 *Tariqilabeo latius*, RGUMF-70, 160 mm SL: lateral view.

Cyprinus latius Hamilton 1822: 345, 393; type locality: Teesta River near Kalimpong Duars and Siliguri Terai, West Bengal, India (26.794276°N, 88.561392°E). [Original type locality: Teesta River, at base of Darjeeling Himalayas, India/Bangladesh.]

Tariqilabeo latius: Kottelat, 2016: 445 (valid).

Material examined. RGUMF-70–73, 18 exs., 65.0–172 mm SL, Panye River, Ranga River at Yazali; Poma River at Ramgath and Ranga River at Kimin in Papum Pare district, Arunachal Pradesh.

Vernacular names in Arunachal Pradesh. *Ngoku-ngoi* (in Nyishi), *Ngopui* (in Adi), *Yirsoo* (in Galo).

Diagnosis. Snout rounded, and lacking rostral flaps and tubercles. Mouth inferior, upper lip thin, wide at the corner of mouth to connect the lower jaw, lower lip connected posteriorly to underside of head without mental disc. Lateral line scales with 34–40 plus 1–4 pored scales on caudal fin, predorsal scales 8–11, circumpeduncular scales 16–24, and vertebrae 35–38. Eyes large with diameter 24.4%–38.1% HL.

Coloration. In live, upper half of body yellowish brown to dark brown, lower half yellowish. Flanks and lateral portions of head mottled dark brown, with irregular clusters of black spots. Larger specimens with dark humeral spot on 5th–6th lateral-line scale. Ventral surface of body is cream to yellow. Dorsal and caudal fins yellow-grayish, other fins orange.

Distribution. Known from the Ganga-Brahmaputra basin in India, Nepal, Bhutan, and Bangladesh. In Arunachal Pradesh: abundantly found in streams and tributaries of Siang, Dibang, Kameng, Subansiri, Ranga, and Dikrong rivers.

Remarks. Catch per unit of this species is highest in most of the major rivers of Arunachal Pradesh.

Economic importance. A food fish.

Conservational status: A widely spread species occurring throughout the Ganga-Brahmaputra river basin. IUCN status: Least concern (LC).

22. Genus: *Garra* Hamilton, 1822

Garra (subgenus of *Cyprinus*) Hamilton 1822: 343, 393. Type species: *Cyprinus (Garra) lamta* Hamilton 1822. Type by subsequent designation.

Body cylindrical; head slightly depressed; blunt snout, heavily tuberculated or absent, transverse groove present or absent; mouth inferior, semicircular; lips thick and fleshy; upper lip fimbriated; lower jaw covered by thick labial fold; lower lip with a mental adhesive disc consisting of semicartilaginous pad.

Key to the species of *Garra*:

1a. Snout with both proboscis and transverse lobe or only transverse lobe……..……..2
1b. Snout lacking both proboscis and transverse lobe (=Snout smooth)………………..10
2a. Snout with both proboscis and transverse lobe ………………………..3
2b. Snout with only transverse lobe …………………….. …………… .7
3a. Proboscis lacking prominent lobes ………………………………… 4
3b. Proboscis with 2–3 prominent large lobes………………………………5
4a. 10–12 branched pectoral-fin rays; lateral line scale rows 32–33; circum-peduncular scale rows 16………………………………….. *G. kalpangi*
4b. 14–16 branched pectoral-fin rays; lateral line scale rows 35–36; circum-peduncular scale rows 12……………………………..… *G. nigricauda*
5a. Lateral line scales 35 and above ……………………………….…… *G. arunachalensis*
5b. Lateral line scales 34 and below ……………………………………6
6a. Proboscis with 2 prominent lobes; transverse scale rows between the lateral line and the pelvic-fin origin 3½…………………………*G. birostris*
6b. Proboscis with 3 prominent lobes; transverse scale rows between the lateral line and the pelvic-fin origin 2½……………………… …*G. tamangi*
7a. Lateral line scales fewer than 35…………………………………….. *G. lamta*
7b. Lateral line scales 35 and above …………………………………8
8a. A faint black to dusky blotch at the caudal-fin base on the margin of hypural complex…………………………………….……*G. magnidiscus*
8b. No faint blackish or dusky blotch at the caudal-fin base……………….……..9
9a. 39 lateral line scales……………………………………..…*G. kempi*
9b. 35–36 lateral line scales ………………………………………*G. arupi*
10a. Caudal fin with a "W"-shaped band………………………………..…*G. lissorhynchus*
10b. No "W"-shaped band on caudal fin……………………………………..…11
11a. Maximum adult size ca. 53.1 mm SL, 14–16 branched pectoral fin rays……………………………………….*G. minima*
11b. Adult size greater than 60 mm SL, branched pectoral fin rays fewer than 13……………………………………………………..12

12a. A bluish spot behind opercle and 5–6 midlateral reddish spots on anterior parts of body..*G. annandalei*
12b. No spot behind opercle and flank...13
13a. Nine scale rows between the dorsal and ventral fins...................
.........*G. rupicola*
13b. Scale rows between the dorsal fin and the ventral fin fewer than 9............14
14a. Subcutaneous breast scales, 35–36 lateral line scales.................. *G. alticaputus*
14b. Naked breast, 33–34 lateral line scales.............................. *G. kimini*

***Garra annandalei* Hora 1921**
Fig. 42.

FIG. 42 *Garra annandalei*, RGUMF-75, 78.3 mm SL: lateral view.

Garra annandalei: Hora, 1921: 657; type locality: Assam and various streams at the base of the Darjiling [Darjeeling] Himalayas, West Bengal, India.

Garra annandalei: Menon, 1999: 142 (valid).

Material examined. RGUMF-74–76, 29 exs., 35.0–100.0 mm SL, Ranga River at Kimin, Panye River in Papum Pare district and Kameng at Bhalukpong in West Kameng district, Arunachal Pradesh (AP). RGUMF-unregistered, 05 exs., 65.2–98.7 mm SL, Lohit River at Tezu, Lohit district, AP.

Vernacular names in Arunachal Pradesh. *Ngop/Ngoha* (in Nyishi), *Ngoyu* (in Galo), *Ngopui* (in Adi).

Diagnosis. Snout smooth with a shallow transverse groove, 34–38 scales on lateral line, 8 series of scale rows from base of dorsal fin to origin of ventral fin. Prominent scales on belly and inconspicuous on chest.

Distribution. India: Assam, Darjeeling Himalayas, and Arunachal Pradesh. Also known from Kosi drainage of eastern Nepal. In Arunachal Pradesh: see material examined.

Coloration. In live, some of the scales along the lateral line pinkish to reddish. Body and head dark gray ventrally whitish, a black spot (bluish in live) behind the upper angle of gill opening.

Remarks. Menon (1964) mentioned that *G. annandalei* lacks the transverse groove on snout but the original description of Hora (1921) clearly described the presence of a transverse groove on snout.

Economic importance. This species is a locally popular food fish.

Conservational status: This species is widely distributed throughout the Ganga-Brahmaputra basin. IUCN status: Least concern (LC).

***Garra lissorhynchus* (McClelland 1842)**

Fig. 43.

FIG. 43 *Garra lissorhynchus*, RGUMF-373, 88.3 mm SL: (A) lateral, and (B) ventral views.

Platycara lissorhynchus McClelland 1842: 587, pl. 18, Fig. 2; type locality: Kasyah mountains (= Khasi Hills), Meghalaya, India.

Garra lissorhynchus: Vishwanath et al., 2007: 86 (valid).

Materials examined. RGUMF-81, 4 exs., 38–70 mm SL, Noadhing at Namdapha, Tirap district, Arunachal Pradesh (AP). RGUMF-373, 14 exs., 45.4–92.1 mm SL, Kameng River at Tippi, West Kameng district, AP.

Vernacular names in Arunachal Pradesh. *Ngop/Ngoha* (in Nyishi), *Ngoyu* (in Galo), *Ngopui* (in Adi).

Diagnosis. Snout: smooth and rounded, lacking transverse groove and proboscis, interorbital region somewhat convex; outermost gill arch with 7–9 rakers on the lower arm; scales absent in chest and belly; dorsal fin with a submarginal blackish transverse streak and a "W"-shaped band on caudal fin.

Coloration. In 10% formalin, dorsal and lateral surface of body brownish, ventrally whitish. A black spot (bluish in live specimens) behind the upper angle of the gill opening and also a faint blackish blotch at the base of caudal fin present. Dorsal fin with a submarginal blackish transverse streak and a "W"-shaped band on caudal fin.

Distribution. Known from Meghalaya, Assam, and Arunachal Pradesh and also from the Barak basin in Manipur. In Arunachal Pradesh: Noadhing River, Subansiri River, and Dikrong River.

Remarks. In the male, the color of the body is much lighter and the characteristic markings on the dorsal and the caudal fins are faint (Hora, 1921).

Economic importance. This species is locally popular food fish.

Conservational status: This species is widely distributed in the Brahmaputra and Meghana basins in the Northeast India. IUCN status: Least concern (LC).

***Garra kalpangi* Nebeshwar, Bagra & Das 2012**

Fig. 44.

FIG. 44 *Garra kalpangi*, RGUMF-06, holotype, 60.0mm SL: (A) lateral, and (B) mental adhesive disc.

Garra kalpangi, Nebeshwar et al., 2012: 2355, Figs. 1–2; type locality: River at Yachuli (Brahmaputra River system), 27°25′54″N, 93°46′42″E, elevation 843 m, Lower Subansiri District, Arunachal Pradesh, India.

Garra kalpangi: Shangningam & Vishwanath, 2015: 272 (valid).

Materials examined. RGUMF-06, holotype, 60.0 mm SL, Kalpangi River at Yachuli, Lower Subansiri district, Arunachal Pradesh, India, collected by Kenjum Bagra, July 18, 2005. RGUMF-07, paratype, 50.0–72.4 mm SL, 9 exs., collection data same as above.

Vernacular names in Arunachal Pradesh. *Ngop/Ngoha* (in Nyishi), *Ngoyu* (in Galo), *Ngopui* (in Adi).

Diagnosis. Proboscis squarish shaped, poorly developed. Mental disc elliptical and its length shorter than width. Lateral line complete with 32–33 scales, transverse scales row 3½/1/3½–4, circumpeduncular scales 16, and predorsal scales 10–11. Total vertebrae 31.

Coloration. Dorsal and lateral parts of head and body gray, ventrally yellowish. Dorsal, anal, pelvic, and pectoral fins grayish-white, caudal fin lighter in color with a distal marginal stripe on upper and lower lobe. A black spot behind the upper angle of gill opening is present.

Distribution. The species is presently known only from the Kalpangi River in Lower Subansiri District (Brahmaputra basin), Arunachal Pradesh.

Economic importance. This species is locally popular food fish.

Conservational status: Currently, regarded as a species restricted in distribution only in the Subansiri River drainage (Brahmaputra basin). IUCN status: Not assessed (NA).

***Garra lamta* (Hamilton 1822)**

Fig. 45.

FIG. 45 *Garra lamta*, RGUMF-430, 57.7 mm SL: lateral view.

Cyprinus lamta Hamilton, 1822: 343, 393; type locality: Behar (Bihar) Province and Rapti River, Gorakhpur District, Uttar Pradesh, India.

Garra lamta: Nebeshwar & Vishwanath, 2015: 319 (valid).

Material examined. RGUMF-430, 01 ex., 57.7 mm SL, Dikrong river at Doimukh, Papum Pare district, Arunachal Pradesh.

Vernacular names in Arunachal Pradesh. *Ngop/Ngoha* (in Nyishi), *Ngoyu* (in Galo), *Ngopui* (in Adi).

Diagnosis. Snout rounded, transverse lobe at snout tip demarcated behind by a deep transverse groove, and lateral surface of snout in front of nostrils covered by horny tubercles. First gill arch with 9–14 rakers on lower arm. Mental disc length 1.36 times its width. Scales on lateral line 31–34 and predorsal 8–10. Chest and belly scaled but sparse on chest. Caudal fin margin emarginate. A faint spot near the base of the caudal fin present.

Coloration. Head and body dark gray above with greenish reflection, ventrally dull white to creamish. A faint blackish spot at the base of caudal fin present.

Distribution. Known from the Ganga Brahmaputra drainage in Sikkim, Darjeeling, Kumaon Himalayas, and eastern Nepal (Menon, 1964). In Arunachal Pradesh: see material examined.

Remarks. Hamilton's (1822) description of *G. lamta* is inadequate to distinguish from its congeners. Menon (1964) redescribed this species based on the specimens collected from the type locality of the species. However, Nebeshwar and Vishwanath (2015) disagree with Menon's (1964) description of *G. lamta* on the fact that the presence of a transverse groove and a patch of horny tubercles on sides of snout were not mentioned in the original description. A proper taxonomic study is required for this species. Here, we follow Menon (1964) for the identification of this species, as it is the only detailed reference available till date.

Economic importance. This species is locally popular food fish.

Conservational status: This species is widely distributed throughout the Ganga-Brahmaputra river basin. IUCN status: Least concern (LC).

***Garra kempi* Hora 1921**

Fig. 46.

FIG. 46 *Garra kempi*, RGUMF-438, 85.2 mm SL: lateral view.

Garra kempi Hora, 1921b: 665, pl. 14, Fig. 3 and 3a; type locality: Siyom R., below Damda, at altitude 1300 ft., among Abor Hills (West Siang district, Arunachal Pradesh), India.

Garra kempi: *Bagra* et al., 2009: 340 (listed, valid).

Material examined. RGUMF-438, 04 exs., 60.0–95.0 mm SL, Yomgo River at Mechuka, head water of Siyom river in West Siang district, Arunachal Pradesh.

Vernacular names in Arunachal Pradesh. *Ngop/Ngoha* (in Nyishi), *Ngoyu* (in Galo), *Ngopui* (in Adi).

Diagnosis. Snout rounded, proboscis absent, lateral surface of snout in front of the nostrils slightly raised from the general surface; scales absent on chest and sparsely present along midline of belly; pelvic fin not reaching anal origin, 39 lateral line scales, 8 rows of scales between the bases of the dorsal and ventral fins. Mental disc width 1.46 times its length.

Coloration. Dorsal and lateral surface of body blackish, ventrally dull white. Dorsal and caudal fin grayish. Dorsal parts of paired fins grayish, ventrally dull.

Distribution. Known from the Brahmaputra and Barak basin in Northeast India: Arunachal Pradesh, Manipur, Meghalaya, Mizoram, and Nagaland.

Remarks. This species is encountered up to 1932 msl in Yomgo River at Mechuka (upper reaches of the type locality "Siyom River") during warmer season (March–April). This species is also reported from Tibet (China). However, its occurrence in Tibet is unlikely as the course of the Siang river between Arunachal Pradesh and Tibet plateau have some (at least three) high water falls barriers.

Economic importance. This species is a locally popular food fish.

Conservational status: A widely spread species occurring in the Brahmaputra and Barak basins. IUCN status: Least concern (LC).

***Garra rupicola* (McClelland 1839)**

Fig. 47.

FIG. 47 *Garra rupicola*, RGUMF-346, 71.7 mm SL: (A) lateral, (B) ventral view of belly and mental adhesive disc, and (C) dorsal view of head.

Gonorhynchus rupicolus McClelland, 1839: 281, 373, pl. 43, Figs. 4 and 5; type locality: Lareeh River, few miles beyond Bramacuna [Brahmakund], in Mishmee Hills, Arunachal Pradesh, India, 27°45′N, 96°20′E, elevation 1000 ft.

Garra rupicola: Kottelat, 2013: 103 (valid).

Material examined. RGUMF-346–359, 13 exs., 43.9–77.4mm SL, Brahmaputra River near Brahmakund, Lohit district, Arunachal Pradesh.

Vernacular names in Arunachal Pradesh. *Ngop/Ngoha* (in Nyishi), *Ngoyu* (in Galo), *Ngopui* (in Adi).

Diagnosis. Snout broad, smooth, and depressed. A row of open pores extending round the snout and between the nostrils. Lateral line with 35 scales and 9 scale rows between the dorsal and the ventral fin (McClelland, 1839).

Coloration. Body dark brown above without spots, yellowish white below.

Distribution. Known only from Lareeh River, near Brahmakund, Arunachal Pradesh, India.

Remarks. McClelland (1839) described this species from the Lareeh River, a stream not far from Parashuram Kund (a pilgrimage site) in Arunachal Pradesh. The identity of this species is poorly understood. This species may not have a "W"-shaped band on the caudal fin, as the original description and figure does not provide any clue for such band. *Garra abhoyai* Hora 1921 (currently a valid species), which has a "W"-shaped band on caudal fin, was once treated as a synonym of *G. rupicola*. Thereafter, many authors still use the coloration of *G. abhoyai* as a characteristic color feature of *G. rupicola*. One or more presently considered valid species of *Garra* might be a synonym of this species.

Economic importance. This species is a locally popular food fish.

Conservational status: The identity of this species is poorly understood. The presently known distribution of this species is from its type locality only. IUCN status: Not assessed (NA).

***Garra magnidiscus* Tamang 2013**
Fig. 48.

FIG. 48 *Garra magnidiscus*, ZSI/V/APFS/P-623, paratype, 82.7mm SL: (A) lateral, (B) dorsal view of head, and (C) ventral view of mental adhesive disc.

Garra magnidiscus Tamang, 2013: 32, Figs. 1 and 2; type locality: a tributary of the Siang River, about 3 km from Bomdo Village on main road to Tuting, Upper Siang district, Arunachal Pradesh, India.

Garra magnidiscus: Gurumayum & Kosygin, 2016: 90 (valid).

Material examined. ZSI/V/APFS/P-623, paratype, 82.7 mm SL; India: Arunachal Pradesh: Upper Siang district: a tributary of Siang River, about 3 km from Bomdo village on main road to Tuting, 28°44.04′N 94°51.97′E, 429 m asl; coll. Lakpa Tamang, October 26, 2011.

ZSI/V/APFS/P-623, 12, paratypes, 52.7–82.7 mm SL, same collection data as above.

Vernacular names in Arunachal Pradesh. *Ngop-tatum* (in Nyishi), *Ngoyu* (in Galo), *Ngopui* (in Adi).

Diagnosis. Snout broadly rounded, anterior rim of snout provided with a transverse lobe bearing about 7–9 small unicuspid tubercles arranged in a series. Transverse groove discontinuous medially, shallow furrow extend only above rostral barbels to lateral extremity of rostral fold. Proboscis absent. Mental adhesive disc: very large, the posterior margin of disc reaching to or very close to an imaginary transverse line drawn through the origin of left and right pectoral fin; 40–42 perforated lateral line scales, faint blackish blotch at caudal-fin base.

Coloration. In live, dorsal part of head and body blackish brown, lighter below. Ventral part of head and body dull white. Eye pupil black with orange ring border. Caudal-fin base with a faint blackish blotch. A faint dark to dusky streak on outer margin of each caudal-fin lobe and on median rays; upper streak narrower than lower and median rays broader than both.

Distribution. The species is presently known only from the Siang River drainage in Arunachal Pradesh.

Remarks. This species differs from all known *Garra* species of Arunachal Pradesh by its extraordinarily large mental adhesive disc.

Economic importance. This species is a locally popular food fish.

Conservational status: Currently, regarded as a species restricted in distribution only in the Siang River drainage (Brahmaputra basin). IUCN status: Not assessed (NA).

***Garra arupi* Nebeshwar et al. 2009**

Fig. 49.

FIG. 49 *Garra arupi*, RGUMF-unregistered; 71.5 mm SL: lateral view.

Garra arupi Nebeshwar et al., 2009: 198, Figs. 1, 2, and 3; type locality: Deopani River at Roing, Lower Dibang Valley, Arunachal Pradesh, India.

Garra arupi: Gurumayum & Kosygin (2016): 90 (valid).

Materials examined. RGUMF-184; holotype, 60.0 mm SL, Deopani River at Roing, Lower Dibang Valley, Arunachal Pradesh, India. RGUMF-185, 15 exs., 50.0–72.4 mm SL, same data as holotype; RGUMF-186, paratype, 10 exs., 36.5–69.0 mm SL, Iphipani River at Roing, Lower Dibang Valley district. RGUMF-187, paratype, 12 exs., 35.5–65.5 mm SL, Sisiri River at Dunbuk, Lower Dibang Valley; RGUMF-188, paratype, 12 exs., 39.5–65.8 mm SL, Lohit River at Tezu, Lohit district. RGUMF-unregistered, 10 exs., 67.2–72.0 mm SL, Kameng River at Tippi, West Kameng district, AP.

Vernacular names in Arunachal Pradesh. *Ngop-tatum* (in Nyishi), *Ngoyu* (in Galo), *Ngopui* (in Adi).

Diagnosis. Snout rounded with a shallow transverse groove across it. Transverse lobe bears 6–9 pointed horny tubercles and 3–5 small pointed tubercles of the lateral surface of snout. Proboscis poorly developed. Dorsal fin with a transverse subdistal grayish streak and the medial caudal-fin rays from 9th to 13th (counting from upper to lower) principal rays dark gray. Scales in transverse rows 3½–4½/1/2½, scales from lateral line to anal origin 3½–4½; circumpeduncular scales 15–16, predorsal 11–12 scales, and lateral line complete with 35–36 scales. No "W" shaped band in caudal fin. Largest adult size ca. 72.2 mm SL.

Coloration. In 10% formalin, dorsal parts of head and body dark brown, ventral part of head and abdomen whitish or yellowish. A blackish submarginal band on dorsal fin. A small rounded black spot at the upper angle of gill opening (in live, this black spot is bluish), medial rays of caudal fin blackish.

Distribution. Known only from the Brahmaputra basin in Arunachal Pradesh: the Iphipani, Deopani, and Sisiri rivers of Lower Dibang Valley district (tributaries of Dibang River); from Lohit River in Lohit district and also from Kameng River in West Kameng district.

Economic importance. This species is a locally popular food fish.

Conservational status: This species is known from many major rivers in Arunachal Pradesh (see distribution section). IUCN status: Not assessed (NA).

***Garra arunachalensis* Nebeshwar and Vishwanath 2013**

Fig. 50.

Garra arunachalensis Nebeshwar & Vishwanath, 2013: 101, Fig. 3; type locality: Lower Divang [Dibang] valley district, Deopani River at Roing, Brahmaputra basin, 29°09′35″N, 95°54′08″E, Arunachal Pradesh, India.

Garra arunachalensis: Gurumayum & Kosygin (2016): 92 (valid).

Materials examined. RGUMF-190/5, paratypes, 5 exs., 93.5–128.0 mm SL; Arunachal Pradesh (AP): East Kameng district: Pachabong River at Seppa (Brahmaputra basin), 27°16′35″N 92°53′16″E; K. Bagra and party, February

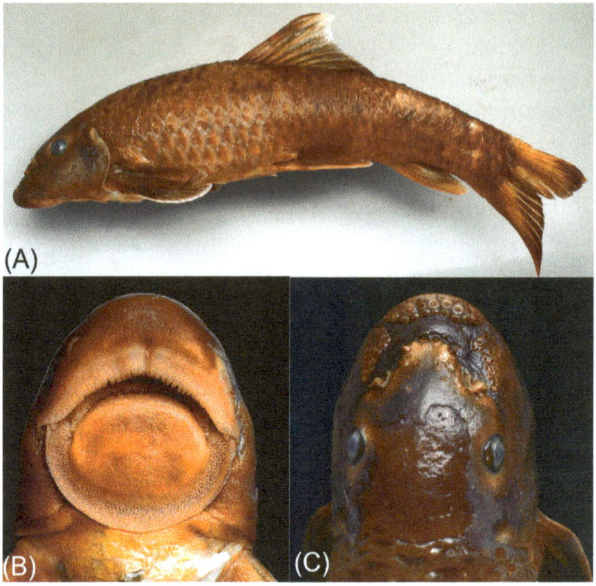

FIG. 50 *Garra arunachalensis*, RGUMF-190, 128 mm SL: (A) lateral view, (B) mental adhesive disc, and (C) dorsal view of head.

13, 2007. RGUMF-unregistered, 15 exs., 105–120 mm SL; Sinyot River at Payum (Siang drainage), West Siang district, AP (Brahmaputra basin).

Vernacular names in Arunachal Pradesh. *Ngop-tatum* (in Nyishi), *Ngoyu* (in Galo), *Ngopui* (in Adi).

Diagnosis. Snout: broadly rounded, anterior rim of snout provided with a transverse lobe bearing about 8–24 unicuspid tubercles, a moderately wide shallow transverse groove separating the transverse lobe from proboscis and lateral surface. Proboscis: moderately elevated from rest of the snout region, more or less quadrate or square in shape, and both the anterolateral corner bears a large horn-like tubercle and another smaller tubercle in between. Lower lip lacks anterolateral lobe. Lateral line complete with 35 scales, 4–4½/1/3½ transverse scale rows, 10–12 predorsal scales, and vertebrae 33–34.

Coloration. In 10% formalin, head, dorsum, and flank brownish gray. Ventral parts of body and head, abdomen yellowish white.

Distribution. Currently known from three major river drainages of the state Arunachal Pradesh in Northeast India: the Dibang, Siang, and Kameng Rivers (all Brahmaputra basins).

Remarks. This species is comparatively larger than other species of *Garra*. Maximum recorded size is ca. 140 mm SL.

Economic importance. This species is a locally popular food fish.

Conservational status: This species is widely distributed in Arunachal Pradesh. IUCN status: Not assessed (NA).

Garra birostris Nebeshwar and Vishwanath 2013
Fig. 51.

FIG. 51 *Garra birostris*, RGUMF-78, paratype, 143 mm SL: (A) lateral view (B) mental adhesive disc, and (C) dorsal view of head.

Garra birostris Nebeshwar & Vishwanath: 2013: 104, Fig. 5; type locality: Papum Pare district, Dikrong River at Doimukh, Brahmaputra basin, 27°08′19″N, 93°44′51″E, Arunachal Pradesh, India.

Garra birostris: Gurumayum & Kosygin (2016): 92 (valid).

Materials examined. RGUMF-77, 5 exs., 60.0–120.0 mm SL; India: Arunachal Pradesh: Papum Pare district: Ranganadi River at Kimin, 27°21′01″N 93°57′11″E; K. Bagra, June 4, 2005. RGUMF-78, 5 exs., 65.0–143.0 mm SL; India: Arunachal Pradesh: West Siang district: Sipu River at Doji (a tributary of Yomgo River, Siang River drainage); K. Bagra, July 17, 2005. RGUMF-80, 5 exs., 40.0–85.0 mm SL; Arunachal Pradesh: Papum Pare district: Poma River at Ramgath, 26°57′38″N 93°24′26″E; K. Bagra, March 7, 2005.

Vernacular names in Arunachal Pradesh. *Ngop-tatum* (in Nyishi), *Ngoyu* (in Galo), *Ngopui* (in Adi).

Diagnosis. Snout: moderately rounded, anterior rim of snout coincide with a transverse lobe bearing about 11–19 uni to tetra cuspid tubercles, and transverse lobe well separated from the proboscis and lateral surface by a deep transverse groove. Proboscis bilobed and each lobe bears a tri or tetra cuspid acanthoid tubercle. Lateral line complete with 33–34 scales, 4½/1/3½ lateral transverse scale rows, 10–11 predorsal scales, circumpeduncular scale rows 16, and total vertebrae 32.

Coloration. In 10% formalin, head, dorsum, and flank dark brown or dark gray. Ventral parts of head and body yellowish white or light gray A black spot at upper angle of gill opening.

Distribution. The species is known from three head waters of the Brahmaputra basin in Arunachal Pradesh: Kameng River, Sipu River, and tributaries of Yomgo River (Siang River drainage) and Dikrong River.

Remarks. No transverse groove form in juvenile specimen of ca. 64.0mm SL, and also proboscis very poorly developed.

Economic importance. This fish is a locally popular food fish.

Conservational status: This species is widely distributed in Arunachal Pradesh. IUCN status: Not assessed (NA).

***Garra tamangi* Gurumayum and Kosygin 2016**

Fig. 52.

FIG. 52 *Garra tamangi*, ZSI/APRC/P-1175, holotype, 153.9mm SL: (A) lateral view (B) mental adhesive disc, and (C) dorsal view of head.

Garra tamangi: Gurumayum & Kosygin, 2016: 85, Figs. 1 and 2; type locality: Dikrong River at Hoj near NHPC Hydel complex, a tributary of the Brahmaputra River basin, Papum Pare district, Arunachal Pradesh, India.

Material examined. ZSI/APRC/P-1175, holotype, 153.9mm SL, India, Arunachal Pradesh, Dikrong River at Hoj near NHPC Hydel complex, a tributary of Brahmaputra River basin, Papum Pare district, collectors: S.D. Gurumayum and L. Tamang, 24 March 2015.

ZSI FF 5423, 2 exs., paratype, 82.0–104.6mm SL; ZSI/APRC/P-1176, paratype, 3 exs., 67.9–153.9mm SL, same data as holotype.

Vernacular names in Arunachal Pradesh. *Ngop-tatum* (in Nyishi), *Ngoyu* (in Galo), *Ngopui* (in Adi).

Diagnosis. Snout moderately rounded with well-developed transverse lobe. Transverse lobe covered with 15–25 uni to tetra cuspid tubercles and posteriorly demarcated by a transverse groove. Proboscis roughly triangular in dorsal view, trilobed anteriorly with one lobe each on the anterolateral corner and another lobe in between. Mental disc elliptical, shorter than wide.

Coloration. In 10% formalin, head, dorsum, and side dark brownish dark gray. Ventral parts of body and head pale yellowish. A faint blackish spot immediately anterior to upper angle of gill opening.

Distribution. Known only from Dikrong River in Papum Pare district, Arunachal Pradesh.

Economic importance. This fish is a locally popular food fish.

Conservational status: Currently, regarded as a species restricted in distribution only in Dikrong River drainage (Brahmaputra basin). IUCN status: Not assessed (NA).

Garra minima Arunachalam et al., 2013

Fig. 53.

FIG. 53 *Garra minima*, RGUMF-419, 57.5 mm SL: (A) lateral view, (B) dorsal view of snout, and (C) mental adhesive disc.

Garra minimus, Arunachalam et al., 2013: 124, Fig. 4, type locality: tributary of Ranga River, Lower Subansiri district, Arunachal Pradesh, India, 27°21′35.9″N, 93°42′21.4″E.

Garra minima: Arunachalam & Nandagopal, 2014: 49 (valid).

Material examined. RGUMF-419, 03 exs., 54.5–57.5 mm SL, Ranganadi River at Kimin, Papum Pare district, Arunachal Pradesh.

Vernacular names in Arunachal Pradesh. *Ngop/Ngoha* (in Nyishi), *Ngoyu* (in Galo), *Ngopui* (in Adi).

Diagnosis. Not well-developed proboscis and shallow transverse groove, belly scaled, and chest naked, predorsal scale with 10–11, lateral line scale with 35–37, transverse scale row between the dorsal-fin base and anal-fin base 2.5–3.5/3, adult size grow up to 53.1 mm SL.

Coloration. In 10% formalin, dorsal and lateral surface of head and body dark brown to grayish, whitish beneath. Dorsal fin with a transverse subdistal grayish streak, bases of dorsal-fin rays blackish; no markings on the pectoral, pelvic, and anal fins, median caudal-fin rays grayish.

Distribution. Known only from the Ranga River in Arunachal Pradesh.

Remarks. Specific name was originally "*minimus*," but corrected later into its feminine adjective as "*minima*."

Economic importance. This fish is a locally popular food fish.

Conservational status: Currently, regarded as a species restricted in distribution only in Ranganadi River drainage (Brahmaputra basin). IUCN status: Not assessed (NA).

Garra alticaputus

Fig. 54.

FIG. 54 *Garra alticaputus*, RGUMF-418, 51.7 mm SL: (A) lateral view, (B) dorsal view of snout, and (C) mental adhesive disc.

Garra alticaputus, Arunachalam et al., 2013: 128, Fig. 6, type locality: Dikrong River at Boorum Village, tributary of Ranga River, Lower Subansiri District, Arunachal Pradesh, India, 27°09′26″N, 93°42′22.7″E.

Garra alticaputus: Gurumayum & Kosygin 2016: 90 (valid).

Material examined. RGUMF-418, 01 ex., 51.7 mm SL, Ranganadi river at Kimin, Papum Pare district, Arunachal Pradesh.

Vernacular names in Arunachal Pradesh. *Ngop/Ngoha* (in Nyishi), *Ngoyu* (in Galo), *Ngopui* (in Adi).

Diagnosis. Not well developed proboscis and shallow transverse groove, belly scaled and subcutaneous scales in chest region, 9–10 predorsal scales, 35–36 lateral line scales, transverse scale rows 4.5/3.0–3.5.

Coloration. In live, dorsal and lateral surface of head and body brownish to gray, dirty white beneath. Dorsal fin base yellowish, subdistally dusky; no markings on pectoral, pelvic and anal fins; median caudal-fin rays darker than rest of the fin.

Distribution. Known only from the Ranga and Dikrong rivers in Arunachal Pradesh.

Economic importance. This species is a locally popular food fish.

Conservational status: Currently, regarded as a species restricted in distribution only in Ranganadi River drainage (Brahmaputra basin). IUCN status: Not assessed (NA).

Garra nigricauda

Fig. 55.

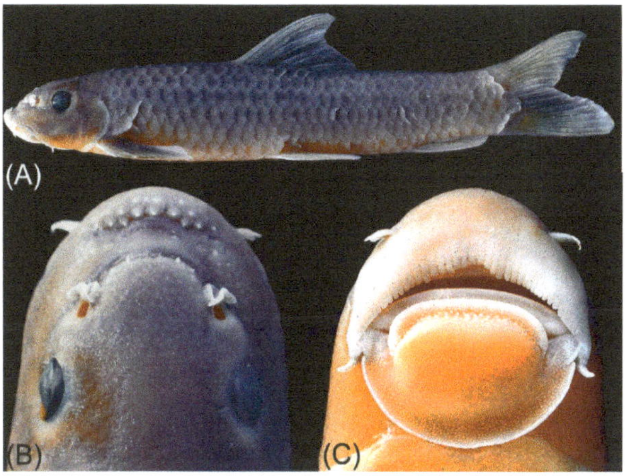

FIG. 55 *Garra nigricauda*, RGUMF-420, 127 mm SL: (A) lateral view, (B) dorsal view of snout, and (C) mental adhesive disc.

Garra nigricauda, Arunachalam et al., 2013: 131, Fig. 8, type locality: Siang River near Pagsighat [Pasighat], Arunachal Pradesh, India, 27°55′22″N, 95°05′54″E.

Garra nigricauda: Gurumayum & Kosygin, 2016: 90 (valid).

Material examined. RGUMF-420, 01 ex., 127.8 mm SL, Siang River at Pasighat, East Siang district, Arunachal Pradesh.

Vernacular names in Arunachal Pradesh. *Ngop-tatum* (in Nyishi), *Ngoyu* (in Galo), *Ngopui* (in Adi).

Diagnosis. Proboscis with slight protrusion and shallow transverse groove; well-developed scales on chest and belly region, 14–16 branched pectoral-fin rays, 10 predorsal scales, 33 lateral line scales, transverse scale rows 4.5/3.5.

Coloration. In 10% formalin, dorsal and lateral surfaces of body brownish-gray, ventrally creamish. Anal fin with a transverse blackish streak, median caudal-fin rays blackish, more prominent on lower side.

Distribution. Known only from Siang River in Arunachal Pradesh.

Economic importance. This fish is a locally popular food fish.

Conservational status: Currently, regarded as a species restricted in distribution only in Siang River drainage (Brahmaputra basin). IUCN status: Not assessed (NA).

***Garra kimini* Arunachalam et al., 2013**

Fig. 56.

FIG. 56 *Garra kimini*, RGUMF-431, 73.6 mm SL: (A) lateral view, (B) mental adhesive disc, and (C) dorsal view of snout.

Garra kimini, Arunachalam et al., 2013: 133, Fig. 10, type locality: tributary of Ranga River, 7 km from Hola camp, Lower Subansiri district, Arunachal Pradesh, India, 27°20′59″N, 93°56′37″E.

Garra kimini: Gurumayum & Kosygin 2016: 90 (valid).

Material examined. RGUMF-431, 01 ex., 73.6 mm SL, Ranganadi River at Kimin, Papum Pare district, Arunachal Pradesh.

Vernacular names in Arunachal Pradesh. *Ngop/Ngoha* (in Nyishi), *Ngoyu* (in Galo), *Ngopui* (in Adi).

Diagnosis. Proboscis with slight protrusion and deep, scales on chest naked, and belly scaled, 9–10 predorsal scales, 33–34 lateral line scales, transverse scale rows 4.5/3.5.

Coloration. In live, dorsal and lateral surface of body grayish, ventrally creamish. A faint blackish spot behind the gill opening, no markings on fins.

Distribution. Known only from the Ranga River in Arunachal Pradesh.

Economic importance. This fish is a locally popular food fish.

Conservational status: Currently, regarded as a species restricted in distribution only in Ranganadi River drainage (Brahmaputra basin). IUCN status: Not assessed (NA).

Subfamily: Cyprininae

23. Genus: *Labeo* Cuvier, 1816

Labeo Cuvier 1816: 194. Type species: *Cyprinus niloticus* Forsskal 1775. Type by subsequent designation.

Adult size large, more or less spindle-shaped body; mouth with pronounced rostral cap; upper lip smooth or crenulate or not papillated; lower lip thick with deep post-labial groove narrowly interrupted at isthmus, dorsal fin with no spine; two barbels: maxillary barbels larger than rostral barbels.

Key to the species of *Labeo*:

1a. 65–80 lateral line scales, scale size small.. *L. Gonious*
1b. 64 or below lateral line scales, scale size moderately large......................2
2a. Blackish triangular marking or blotch around 5–6 lateral line scale 3
2b. No blackish marking around or near 5–6 lateral line scales..................4
3a. 10–12 dorsal branched rays, 40–42 lateral line scales, the post-labial groove separates the lower jaw from isthmus, lips slightly fringed............... ....*L. pangusia*
3b. 9–10 dorsal-fin rays, 37–40 lateral line scales ....*L. bata*
4b. Adult with one pair of barbels...5
4b. Adult with two pairs of barbels...6
5a. Lateral line scales 40–43 ....*L. dyocheilus*
5b. Lateral line scales 37–39...*L. boga*
6a. Live body coloration is blackish above and sides, becoming silvery ventrally, ... *L. rohita*
6b. Live body coloration charcoal black all over the body including fins...... .... *L. calbasu*

***Labeo calbasu* (Hamilton 1822)**
Fig. 57.

FIG. 57 RGUMF-432, 159 mm SL: (A) lateral view, and (B) ventral view of mouth part.

Cyprinus calbasu, Hamilton, 1822: 297, 387, pl. 2, Fig. 33; type locality: rivers and ponds of West Bengal, India.

Labeo calbasu: Kottelat, 2013: 115 (valid).

Materials examined. RGUMF-52, 3 exs., 100–115 mm SL, Ranga River at Kimin, Papum Pare district, Arunachal Pradesh (AP). RGUMF-432, 2 exs., 103.2–159.6 mm SL, Dikrong River at Doimukh, Papum Pare district, AP.

Vernacular names in Arunachal Pradesh. *Ngopii-yalik* (in Nyishi & Galo), *Ngoley* (in Adi).

Diagnosis. Body stout and deep. Snout depressed and pointed, devoid of lateral lobe.

Mouth inferior, upper and lower lips thick and conspicuously fringed. Two pairs of barbells, dorsal fin with ii–iii,12–15 rays. Scales moderately large, 16–18 predorsal scales, lateral line complete with 40–44 scales, and transverse scale row with 9/1/6.

Coloration. In live, blackish-green above, somewhat lighter ventrally. Flank with scarlet spots. All fins blackish except caudal fin, distal tip of the upper lobe of caudal fin whitish.

Distribution. Known from India, Nepal, Bangladesh, and Pakistan. In Arunachal Pradesh: known from Ranga, Dikrong, and Siang rivers.

Economic importance. This species is a popular food fish.

Conservational status: This fish is a widely spread species occurring in the Ganga-Brahmaputra and the Indus River basins. It is a common fish available in many local fish markets in the states of Arunachal Pradesh and Assam. IUCN status: Least concern (LC).

***Labeo gonius* (Hamilton 1822)**

Fig. 58.

FIG. 58 *Labeo gonius*, RGUMF-51, 120 mm SL: lateral view.

Cyprinus gonius Hamilton, 1822: 293, 387, pl 4, Fig. 82; type locality: India: Ganges River basin.

Labeo gonius: Kottelat, 2013: 116 (valid).

Material examined. RGUMF-51, 10 exs., 75–120 mm SL, Ranga at Kimin, Papum Pare district, Arunachal Pradesh.

Vernacular names in Arunachal Pradesh. *Kuri/Ngopii-yalik* (in Nyishi), *Ngopi* (in Galo), *Ngoley* (in Adi).

Diagnosis. Mouth sub-inferior and narrow, lips thick and fringed; one pair each of maxillary and rostral barbel, maxillary shorter than rostral; post-labial groove uninterrupted; dorsal fin with 2 simple and 16–18 branched rays, scales small, 65–80 lateral line scales, 44–57 preanal scales, and transverse scale rows 9/1/13.

Coloration. In live, greenish black on dorsum, becoming dull white on flanks and belly.

Distribution. Known from India, Bangladesh, Nepal, Myanmar, and Pakistan. In Arunachal Pradesh: widely cultured in ponds throughout the state. Wild stocks are also encountered in the Ranga, Dikrong, Noadhing, and Siang rivers.

Economic importance. This fish is a popular food fish.

Conservational status: A widely spread species, introduced elsewhere in many Asian countries for aquaculture. IUCN status: Least concern (LC).

***Labeo pangusia* (Hamilton 1822)**

Fig. 59.

FIG. 59 *Labeo pangusia*, RGUMF-54, 175 mm SL: (A) lateral view, (B) dorsal view of head, and (C) ventral view of mouth part.

Cyprinus pangusia Hamilton, 1822, 285, 386; type locality: River Kosi, India.

Lobeo pangusia, Gurung et al., 2013: 4882 (valid).

Material examined. RGUMF-54, 3 exs., 130–175 mm SL, Dikrong at Doimukh, Papum Pare district, Arunachal Pradesh.

Vernacular names in Arunachal Pradesh. *Ngopii-yalik* (in Nyishi); *Tingir* (in Adi), and *Tiingir* (in Galo).

Diagnosis. No groove on snout, but bears lateral lobes; mouth narrow, lips thick, slightly fimbriated; post-labial groove often uninterrupted, sometimes a muscle flap over isthmus; ii–iii,9–12 dorsal-fin rays; 40–42 lateral line scales; 7½–8½ transverse scale rows between the dorsal fin and lateral line.

Coloration. Body usually dull gray with greenish and reddish tinge. All fins dark gray.

Distribution. This species is native in the Ganga and Brahmaputra drainage system of India, Bhutan, Nepal, and Bangladesh. It is also known from Pakistan. In Arunachal Pradesh: Kameng, Subansiri, Dikrong, Ranga, Dibang, Buridhing, Lohit rivers.

Economic importance. This fish is a popular food fish.

Conservational status: This fish is a widely spread species (see distribution section). However, It is a common fish available in many local fish markets in the states of Arunachal Pradesh and Assam. IUCN status: Near threatened (NT)

***Labeo rohita* (Hamilton 1822)**

Fig. 60.

FIG. 60 *Labeo rohita*, RGUMF-unregistered, 220 mm SL, lateral view.

Cyprinus rohita Hamilton, 1822, 301, pl. 36, Fig. 85; type locality: Gangetic provinces, India.

Labeo rohita: Kottelat, 2001: 58 (valid).

Materials examined. RGUMF-53, 4 exs., 116–125 mm SL, Dikrong at Doimukh, Papum Pare district, Arunachal Pradesh. RGUMF-unregistered, 220 mm SL, Lohit River at Tezu, Lohit district, Arunachal Pradesh.

Vernacular names in Arunachal Pradesh. *Rohu/Ngopii-yalik* (in Nyishi & Galo), *Ngoley* (in Adi).

Diagnosis. Snout without lateral lobe; lips thick and fringed with a distinct inner fold, isthmus rounded, dorsal fin with 13–14 branched rays, 12–16 predorsal scales, and lateral line complete with 40–42 pored scales.

Coloration. In live, bluish to brownish along the back, silvery with tinged red along flank and belly whitish. In most of the specimens, the fins are blackish.

Distribution. Known from northern and central India, Bangladesh, Nepal, Myanmar, and Pakistan. In Arunachal Pradesh: widely cultured in ponds, wild stocks are encountered in Ranga, Dikrong, Lohit, and Siang rivers.

Economic importance. This species is a popular food fish.

Conservational status: A widely spread species, introduced elsewhere in many Asian countries for aquaculture. IUCN status: Least concern (LC).

***Labeo bata* (Hamilton 1822)**
Fig. 61.

FIG. 61 *Labeo bata*, RGUMF-396, 180 mm SL: lateral view.

Cyprinus bata, Hamilton, 1822: 283, 386; type locality: Rivers and ponds of Bengal, India.

Labeo bata: Kottelat, 2013: 115 (valid).

Materials examined. RGUMF-396, 01 ex., 180 mm SL, Lohit River, Lohit district, Arunachal Pradesh.

Vernacular names in Arunachal Pradesh. *Bata Ngopii-yalik* (in Nyishi & Galo), *Ngoley* (in Adi).

Diagnosis. Snout slightly projecting beyond mouth, often studded with pores. Mouth inferior, lips very thin, lower lips slightly fringed with inner cartilaginous covering. A pair of minute maxillary barbel present. Dorsal fin inserted close to snout tip than to caudal-fin base, with ii–iv, 9–10 rays. Scales moderate, lateral line with 37–40 scales.

Coloration. In live, body color above and the dorsal half of flanks is golden yellow, silvery on the lower half and abdomen. An irregular black blotch present on 4th–6th scales of lateral line. Pelvic and anal fin tips orange red.

Distribution. Known from India, Bangladesh, and also reported from Pakistan. In Arunachal Pradesh: Ranga, Dikrong, and Siang rivers.

Economic importance. This species is a popular food fish with a high value in the market.

Conservational status: A widely spread species, commonly available in the local fish market in the states of Arunachal Pradesh, Assam, and West Bengal, India. IUCN status: Least concern (LC).

***Labeo boga* (Hamilton 1822)**
Fig. 62.

FIG. 62 *Labeo boga*, RGUMF-395, 160 mm SL: lateral view.

Cyprinus boga Hamilton, 1822: 286, 386, pl. 28, Fig. 80; type locality: Brahmaputra River, India.

Labeo boga: Kottelat, 2013: 115 (valid).

Material examined. RGUMF-395, 01 ex., 170.0 mm SL, Tengapani River, Namsai district, Arunachal Pradesh.

Vernacular names in Arunachal Pradesh. *Ngopii-yalik* (in Nyishi); *Ngoley* (in Adi).

Diagnosis. Snout moderately projecting beyond jaw, devoid of lateral lobes, covered with small scattered pores; mouth narrow, both lips plain and simple, lower jaw surface covered with a thin cartilaginous layer; maxillary barbel minute, single pair; dorsal fin with 2 simple and 8–9 branched rays; pectoral fin with i,15 rays, its distal tip not extending to pelvic-fin origin. Scales size moderate, lateral line with 37–39 scales, transverse scale rows 6–7½/1/5½, 10–11 predorsal scales, and 20 circumpeduncular scales.

Coloration. In live, body silvery on flanks and orange tinge dorsally; fins light orange with a reddish tinge.

Distribution. Known from India, Nepal, Bangladesh, Pakistan, and Myanmar. In Arunachal Pradesh: see material examined.

Economic importance. This species is a popular food fish.

Conservational status: A widely spread species, commonly available in local fish market in the states of Arunachal Pradesh, Assam, and West Bengal, India. IUCN status: Least concern (LC).

***Labeo dyocheilus* (McClelland 1839)**

Fig. 63.

FIG. 63 *Labeo dyocheilus*, RGUMF-unregistered, 220 mm SL: (A) lateral view, (B) lateral view of head, and (C) ventral view of mouth part.

Cyprinus dyocheilus M'Clelland, 1839: 268, 330, pl. 37, Fig. 1; type locality: Brahmaputra River, India.

Labeo dyocheilus: Kottelat, 2013: 116 (valid).

Material examined. RGUMF-unregistered, 220.0 mm SL, Pare at Kheel, Papum Pare district, Arunachal Pradesh.

Vernacular names in Arunachal Pradesh. *Ngopii-yalik* (in Nyishi); *Tengir* (in Adi and Galo).

Diagnosis. Snout tip blunt with distinct lateral lobe and tuberculated; snout with a prominent groove in adult, which arise as a depression in front of the orbit and extending toward the middle of snout to meet its counterpart on the other side; mouth wide and inferior; lips highly fimbriated with pronounced labial folds; only a pair of maxillary barbel present; dorsal fin with ii–iii,9–10 rays, its origin located almost in the middle point between the snout tip and caudal-fin base; lateral line scales 40–43, and 5½–6½ rows of scales between the pelvic fin and the lateral line, and circumpeduncular scales 19–23.

Distribution. Known from India, Bangladesh; Bhutan, Myanmar and Nepal. In Arunachal Pradesh: Ranga, Dikrong, Siang rivers.

Economic importance. This fish is a popular food fish.

Conservational status: A widely spread species, commonly available in the local fish market in the states of Arunachal Pradesh, Assam and Uttarakhand, India. IUCN status: Least concern (LC).

24. Genus: *Barbonymus* Kottelat 1999

Barbonymus Kottelat 1999: 595. Type species: *Barbus schwanenfeldii* Bleeker 1853. Type by original designation.

Head small, body strongly compressed; dorsal profile elevated, arched, slightly concave above the occiput; snout pointed, mouth terminal; snout with minute tubercles scattered at the anterior edge of lacrimal and they are fewer than 11; orbit width is much larger than the snout length; anterior naris placed on anterior half of orbit; jaws equal, lower lips skin distinctly separated by a groove from the lower jaw; last spinous ray of dorsal fin strongly serrated posteriorly; anal-fin base length slightly shorter than the head length; 4–6 intestine loop, 4–5 interneural, and 9–10 pairs of pleural ribs and rarely 11.

***Barbonymus gonionotus* (Bleeker 1849)**
Fig. 64.

FIG. 64 *Barbonymus gonionotus*, RGUMF-unregistered, 94.6 mm SL, lateral view.

Barbus gonionotus, Bleeker, 1849: 15; type locality: Kalimas River [Brantas], Surabaya, Java, Indonesia.

Barbonymus gonionotus: Kottelat, 2013: 81 (valid).

Material examined. RGUMF-unregistered, 1 ex., 94.6 mm SL, Doimukh, Papum Pare district, Arunachal Pradesh.

Vernacular names in Arunachal Pradesh. *Java Puthi* (in Nyishi & Galo), *Siktar* (in Adi).

Diagnosis. Body strongly compressed, dorsal profile steeply inclined to dorsal-fin origin, head small; snout pointed with terminally placed mouth, snout length slightly shorter than the eye diameter, dorsal fin with 4 spinous and 8 branched rays, last spinous ray of dorsal fin longest and strongly serrated posteriorly with 12–13 serrae; and anal fin with 6–7 branched rays.

Coloration. In live, body silvery white with a golden tint, dorsal and caudal fins grayish yellow; pectoral-fin light yellow, anal and pelvic fins light orange with reddish distal tip.

Distribution. Widely distributed in Cambodia, Indonesia, Lao People's Democratic Republic, Vietnam, Malaysia, and Thailand. In India: introduced elsewhere including the state Arunachal Pradesh in Northeast India.

Remarks. An introduced species for pisciculture.

Economic importance. A food fish.

Conservational status: A widely spread species, introduced elsewhere in many Asian countries for aquaculture. IUCN status: Least concern (LC).

25. Genus: *Cyprinus* Linnaeus, 1758

Cyprinus Linnaeus 1758: 320. Type species: *Cyprinus carpio* Linnaeus 1758. Type by subsequent designation

Body robust and compressed, head relatively small; mouth protrusible, lips thick and fleshy, two pairs of barbels, maxillary pair shorter than the mandibular; pharyngeal teeth in three rows, scales large and thick, caudal fin deeply emarginated.

***Cyprinus carpio* Linnaeus 1758**
Fig. 65.

FIG. 65 *Cyprinus carpio*, RGUMF-unregistered, 180 mm SL: lateral view.

Cyprinus carpio Linnaeus, 1758: 320; type locality: Europe.
Cyprinus carpio: Kottelat, 2013: 94 (valid).

Materials examined. RGUMF-29, 2 exs., 110–115 mm SL, Kele River at Lower Subansiri district, Arunachal Pradesh. RGUMF-unregistered, 180 mm SL, Doimukh bazaar, Papum Pare district, AP.

Vernacular names in Arunachal Pradesh. *Common-carp* (in Nyishi & Galo), *Siktar* (in Adi).

Diagnosis. Mouth subterminal, protrussible with thick and fleshy lips. Barbels two pairs, one each of maxillary and rostral. Scales large, lateral line complete with 33–37 + 2–3, transverse scale rows 5½/1/5½, predorsal scales 10, and circumpeduncular scales 16. Dorsal fin with iv,19 rays, last simple ray strong and serrated posteriorly. Anal fin with two simple and five branched rays, its second simple ray bony and serrated posteriorly.

Coloration. In live, body dark brown or gray dorsally, lighter on flanks, belly whitish, fins dusky.

Distribution. Native to Black, Caspian, and Aral Sea basins. Introduced elsewhere in India including the state Arunachal Pradesh of Northeast India.

Remarks. *Cyprinus carpio* was introduced into Ootacamund, Madras, India in 1939, from Ceylon during the British regime for pisciculture (Chopra, 1951: 45). *C. carpio* and *Carassius carassius* are very similar in morphometrics and meristics but differs from the later in having a pair each of rostral and maxillary barbels (vs. both are absent).

Economic importance. A preferred food fish, and extensively used in rice cum fish culture system in Ziro valley in Arunachal Pradesh.

Conservational status: Introduced elsewhere in many Asian countries for aquaculture. IUCN status: Vulnerable (Vu).

26. Genus: *Carassius* Jarocki, 1822

Carassius (subgenus of *Cyprinus*) Jarocki 1822: 54, 74. Type species: *Cyprinus carassius* Linnaeus 1758. Type by absolute tautonymy.

Body short and moderately compressed; dorsal profile convex, barbels absent; last simple rays of anal and dorsal fins weakly serrated posteriorly; 23–33 gill rakers on the first branchial arch; lateral line with 31–36 scales; golden-green shining body; caudal fin forked.

***Carassius carassius* (Linnaeus 1758)**

Fig. 66.

FIG. 66 *Carassius carassius*, RGUMF-366, 86.6 mm SL: lateral view.

Cyprinus carassius Linnaeus, 1758: 320; type locality: European ponds.

Carassius carassius: Kullander et al., 1999: 108 (description, valid).

Material examined. RGUMF-366, 2 exs., 84.8–86.63 mm SL, Ziro valley, Lower Subansiri district, Arunachal Pradesh.

Vernacular names in Arunachal Pradesh. *Common-carp* (in Nyishi & Galo), *Siktar* (in Adi).

Diagnosis. Body deep and laterally compressed. Mouth subterminal, protrussible with thick lips. Barbels absent. Scales large, lateral line complete with 28–29 scales. Dorsal fin with iv,18 rays, last simple ray strong and serrated posteriorly. Anal fin with three simple and six branched rays, the last simple ray bony and serrated posteriorly.

Coloration. In live, body dark brown or gray dorsally, lighter on flanks, belly whitish, fins dusky.

Distribution. Native to Europe, but widely introduced elsewhere including India.

Remarks. The English Carp, *Carassius carassius*, was introduced in India at Ootacamund, Madras, in 1874, from Europe (Chopra, 1951: 45). The species might be accidentally introduced to Arunachal Pradesh along with the seed of *Cyprinus carpio* as the occurrence of the species is hitherto unknown in the state.

Economic importance. A preferred food fish, and extensively used in rice cum fish culture system in Ziro valley in Arunachal Pradesh.

Conservational status: This species is also introduced elsewhere in many Asian countries for aquaculture. Least concern (LC).

27. Genus: *Chagunius* Smith, 1938

Chagunius Smith 1938: 157. Type species: *Cyprinus chagunio* Hamilton 1822. Type by original designation (also monotypic).

Body and head moderately compressed; abdomen rounded, prominent rostral and maxillary barbels; snout overhanging, divided into a median and two lateral lobes by a groove extending upward and forward from the base of each rostral barbel; suborbital bones narrow; gill openings wide, extending well forward and its membranes narrowly joined to the isthmus; gill rakers in the form of nine triangular plates on lower arm of the first arch; three rows of pharyngeal teeth; snout and cheeks with thick horny tubercles in male, smaller and fewer in females; dorsal fin inserted ahead of pelvic fins; last simple dorsal-fin ray osseous and coarsely serrated posteriorly; anal fin with five branched rays, the last two rays in the adult male are greatly elongated; caudal fin forked; lateral line scales 44–48.

***Chagunius chagunio* (Hamilton 1822)**
Fig. 67.

FIG. 67 *Chagunius chagunio*, RGUMF-404, 190 mm SL: lateral view.

Cyprinus chagunio Hamilton, 1822: 295, 381; type locality: Yamuna River and northern rivers of Bihar and West Bengal, India.

Chagunius chagunio: Menon, 1999: 73 (valid).

Materials examined. RGUMF-43, 4 exs., 75–140 mm SL, Poma River at Poma, Papum Pare district, Arunachal Pradesh (AP). RGUMF-404, 01 ex., 190.0 mm SL, Alubari Ghat, Lohit River, Lohit district, AP.

Vernacular names in Arunachal Pradesh. *Orchin* (in Nyishi & Galo), *Serapeki* (in Adi).

Diagnosis. Mouth subterminal. Body depth greater than lateral head length. Barbel two pairs, both longer than the eye diameter. Snout: truncate, tubercles on sides, and tip smooth. Dorsal fin with iv–v,8 rays, last simple ray osseous and serrated posteriorly. Lateral line scales 45–47, transverse scale rows 11–12/1/7, and 23–25 circumpeduncular scales.

Coloration. In live, back and dorsal parts of head grayish, silvery below.

Distribution. Known from Ganga and Brahmaputra drainages of northern and northeastern India, Nepal, and Bangladesh. In Arunachal Pradesh: Dikrong, Poma, Siang, Dibang, Kameng, Lohit, and Tezu rivers.

Remarks. During breeding season, males developed heavy tubercles on snout, cheek, head, spreading on to dorsum.

Economic importance. This species is a locally popular food fish.

Conservational status: A widely spread species occurring in the Ganga-Brahmaputra basins. IUCN status: Least concern (LC).

28. Genus: *Neolissochilus* Rainboth, 1985

Neolissochilus Rainboth 1985: 26. Type species: *Barbus stracheyi* Day 1871. Type by original designation.

Head broad, body deep anteriorly, snout blunt, mouth subterminal or inferior, lower lip present medially; post-labial groove interrupted in the middle; lips thick; barbels two pairs: a rostral and a longer maxillary; broad infraorbital bone, cheeks (below eye) clustered with numerous tubercles; dorsal fin inserted midway between the tip of snout and base of caudal fin base; last dorsal simple ray osseous, without serration, scales large.

Key to the species of *Neolissochilus*:

1a. Mouth subterminal and truncate, edge of lower jaw sharp, lower labial fold widely interrupted in the middle, cheeks below orbit and snout tip bears numerous tubercles,.. *N. hexagonolepis*
1b. Mouth subterminal and smoothly rounded, edge of lower jaw blunts, lower labial fold continuous by a narrow groove, no tubercles on the cheeks below orbit,..*N. hexastichus*

***Neolissochilus hexagonolepis* (McClelland 1839)**
Fig. 68.

FIG. 68 *Neolissochilus hexagonolepis*, RGUMF-36, 140 mm SL: lateral view.

Barbus hexagonolepis M'Clelland, 1839: 270, 336, pl. 41, Fig. 3; type locality: all large rivers in the Eastern frontier, Assam, India.

Neolissochilus hexagonolepis: Zhang et al., 2016: 91 (valid).

Material examined. RGUMF-36 to 39, 6 exs., 70–150 mm SL, Ranga River (R.) at Kimin Poma R. at Poma in Papum Pare district, Siren R. (tributary of the Siang R.) in East Siang district, Kameng R. at Balukpung in West Kameng district, Arunachal Pradesh.

Vernacular names in Arunachal Pradesh. *Ngoge* (in Nyishi), *Ngasa* (in Nocte), *Gase* (in Galo), *Gao* (in Adi).

Diagnosis. Mouth subterminal and truncate, edge of lower jaw sharp, lower labial fold (behind the lower lip) widely interrupted in the middle, cheeks below orbit and snout tip bears numerous tubercles, snout length equals interorbital space, scale large and hexagonal in shape, transverse scale rows 3½–4/1/3, lateral line complete with 27–30 scales, and predorsal scales 8–11, barbels two pairs, one each of maxillary and rostral, both slightly longer than the eye diameter.

Coloration. In live, scales above lateral line coppery colored with bronze-green at base, and scales below lateral line silvery white. All fins deep slate paling toward fin margin. Juveniles have darker dorsum and a blackish spot at the base of caudal fin.

Distribution. India, Nepal, Bhutan, Myanmar, Thailand, Sumatra (Indonesia), and Peninsular Malaysia. In Arunachal Pradesh: Siang, Ranga, Siren (tributary of Siang R.), Lohit, Poma, Kameng, and Dibang rivers.

Remarks. The fish is commonly known as chocolate mahseer.

Economic importance. This species is a highly valued food and game or sport fish.

Conservational status: This fish is a widely spread species (see distribution section). The fish is occasionally available in some local fish markets in the states of Arunachal Pradesh and Assam. IUCN status: Near threatened (NT).

***Neolissochilus hexastichus* (McClelland 1839)**
Fig. 69.

FIG. 69 *Neolissochilus hexastichus*, RGUMF-405, 135 mm SL: lateral view.

Barbus hexastichus McClelland, 1839: 269, 333, pl. 39, Fig. 2; type locality: great rivers in the plains of India.

Neolissochilus hexastichus: Kottelat, 2013: 127 (valid).

Material examined. RGUMF-405, 135.6 mm SL, Dibang River at Roing, Lower Dibang Valley, Arunachal Pradesh.

Vernacular names in Arunachal Pradesh. *Ngoge* (in Nyishi), *Ngasa* (in Nocte), *Buka pethia/Gao* (in Adi), *Gase* (in Galo).

Diagnosis. Mouth subterminal and smoothly rounded, edge of lower jaw blunts, labial fold (behind the lower lip) continuous by a narrow groove, no tubercles on the cheeks below orbit, snout length shorter than interorbital space, scales large, lateral line complete with 30 scales, and 10 predorsal scales.

Coloration. In live, dorsal parts of head and body greenish gray, reddish yellow below. Distal margin of all fins reddish.

Distribution. Ganga-Brahmaputra basin in the Eastern Himalaya: West Bengal, Sikkim, Assam, and Arunachal Pradesh; Barak basin in Manipur.

Remarks. Record of this species from the Chindwin and Irrawaddy basin of Myanmar need verification.

Economic importance. This species is a highly valued food and game or sport fish.

Conservational status: This fish is a widely spread species (see distribution section). The fish is also occasionally available in some local fish markets in the states of Arunachal Pradesh and Assam. IUCN status: Near threatened (NT).

29. Genus: *Osteobrama* Heckel, 1843

Osteobrama Heckel 1843: 1033. Type species: *Cyprinus cotio* Hamilton 1822. Type by subsequent designation.

Body strongly compressed and fairly deep; abdominal edge trenchant completely or only between the pelvic fin base and to vent; dorsal fin insertion located at a level slightly behind to posterior end of pelvic fin base; anal fin long with 14–36 rays.

***Osteobrama cotio* (Hamilton 1822)**

Fig. 70.

FIG. 70 *Osteobrama cotio*, RGUMF-42, 48.0 mm SL: lateral view.

Cyprinus cotio Hamilton, 1822: 339, 393, pl. 39, Fig. 93; type locality: Bengal, India.

Osteobrama cotio: Vishwanath & Shantakumar, 2007: 2882 (valid).

Materials examined. RGUMF-42, 01 exs., 48.0 mm SL, Poma river at Poma, RGUMF-341, 3 exs., 55.4–64.2 mm SL, Naharlagun fish market, Papum Pare district, Arunachal Pradesh.

Vernacular names in Arunachal Pradesh. *Cotio* (in Nyishi, Nocte, & Galo), *Taga* (in Adi).

Diagnosis. Body very deep, laterally compressed, dorsal profile steeply inclined till forming a hump on back. Scales: small, irregularly arranged, 66 lateral line scales, 14/1/12 transverse scale rows, predorsal with 24–27 scale, circumpeduncular with 28–30 scales. Anal fin with iii,31–32 rays.

Coloration. In live, body bright silvery, dorsum slightly darker due accumulation of fine milanephore pigments on scales. A black spot at dorsal origin and also a blotch behind the operculum on the lateral line present. Fin rays of dorsal, pectoral and pelvic fins light yellowish. Anal fin and distal margin of caudal fin hyaline.

Distribution. Known from the Ganga-Brahmaputra basin in India and Bangladesh. In Arunachal Pradesh: abundantly found in the downstream of Noadhing, Dikrong, Ranga, Subansiri, and Siang rivers.

Remarks. Reports of *Osteobrama cotio* from the peninsular India are unlikely.

Economic importance. It has both food and ornamental value.

Conservational status: A widely spread species occurring in the Ganga-Brahmaputra basins. IUCN status: Not assessed (NA).

30. Genus: *Systomus* McClelland, 1838

Systomus McClelland 1838: 948. Type species: Masc. *Systomus immaculatus* McClelland 1838. Type by subsequent designation.

Adult size large, greater than 80 mm SL; maxillary and rostral barbels present; dorsal fin with iv, 8 rays, its last unbranched ray osseous and strongly serrated; lateral line complete, with 27–34 scales; gill rakers simple, acuminate.

Key to species of *Systomus*:

1a. Height of dorsal is fin equal to or slightly longer than the body depth at dorsal origin, lateral line scales 40–42.. *S. clavatus*
1b. Height of dorsal fin is shorter than the body depth at dorsal origin, lateral line scales 32–33.. 2
2a. 12–13 predorsal scales ....*S. immaculatus*
2b. 9 predorsal scales......................... ....*S. sarana*

***Systomus immaculatus* McClelland 1839**
Fig. 71.

FIG. 71 *Systomus immaculatus*, RGUMF-unregistered, 200 mm SL: lateral view.

Systomus immaculatus, McClelland 1839: 284, 380, pl. 44, Fig. 5; type locality Assam, India.

Systomus immaculatus: Pethiyagoda et al., 2012: 77 (redescription with designation of neotype, valid).

Material examined. RGUMF-unregistered, 01 ex., 200 mm SL, Sinki River, tributary of Siang River, Lower Dibang Valley, Arunachal Pradesh.

Vernacular names in Arunachal Pradesh. *Ngogi* (in Nyishi) *Puthi* (in Galo), *Taga* (in Adi).

Diagnosis. Body deep and compressed. Barbel two pairs: maxillary pair as long as half of the eye diameter and rostral pair shorter. Dorsal fin with 4 unbranched and 8 branched rays, last unbranched ray stiff, strongly serrated posteriorly in its proximal two-thirds. Pectoral fin with 1 unbranched and 15 branched rays. Lateral line complete with 32–33 pored scales, 5½ scales in transverse line between the lateral line and origin of dorsal fin, and 12–13 predorsal scales.

Coloration. In 10% formalin, head and body grayish brown, lighter below.

Distribution. Known only from the Brahmaputra drainage in Assam and Arunachal Pradesh.

Remarks. It is the first record of this species from Arunachal Pradesh.

Economic importance. A popular local food fish.

Conservational status: A widely spread species occurring in the Brahmaputra basin only. IUCN status: Not assessed (NA).

***Systomus sarana* (Hamilton 1822)**

Fig. 72.

FIG. 72 *Systomus sarana*, RGUMF-340, 145 mm SL: lateral view.

Cyprinus sarana Hamilton, 1822: 307, 388; type locality: ponds and rivers of Bengal.

Systomus sarana: Pethiyagoda et al., 2012: 77 (valid).

Materials examined. RGUMF-50, 1 ex., 120 mm SL, Poma River at Poma, Papum Pare district, Arunachal Pradesh. RGUMF-340, 1 ex., 145.8 mm SL, Sinkin River at Anpum, Lower Dibang Valley district, Arunachal Pradesh.

Vernacular names in Arunachal Pradesh. *Gnogi* (in Nyishi & Galo), *Taga* (in Adi).

Diagnosis. Body deep and compressed, barbels two pairs, dorsal-fin spine slender and finely serrated posteriorly, lateral line complete with 32 scales, 9 predorsal scales, 5 transverse scale rows between the dorsal-fin origin and lateral line and 4 between the anal-fin origin and lateral line.

Coloration. Flank and belly silvery with golden reflection, dorsum olivaceous, checks golden, and fins dusky to orange.

Distribution. Widely distributed in India (northern and Northeast India), Bangladesh, Nepal, Bhutan, Pakistan, and Afghanistan. In Arunachal Pradesh: see materials examined.

Remarks. Earlier this species was under the genus *Puntius*. Pethiyagoda *et al.* (2012) recognize this species under the genus *Systomus* McClelland, 1838, a genus which has the following key characteristics: adult size greater than 80 mm standard length, dorsal fin with four simple and eight branched rays, the last (= fourth) simple ray spinous and serrated posteriorly, both maxillary and rostral barbels present, lateral line complete with 27–34 scales, and lack post-epiphyseal fontanelle.

Economic importance. A popular food fish.

Conservational status: A widely spread species occurring in the Ganga-Brahmaputra and Indus River basins. IUCN status: Least concern (LC).

Systomus clavatus (McClelland 1845)
Fig. 73.

FIG. 73 *Systomus clavatus*: lateral view (redrawned from McClelland, 1845, pl. 21, Fig. 2).

Barbus clavatus McClelland, 1845: 280, pl. 21, Fig. 2; type locality: rivers at foot of Sikkim mountains, northern frontier of West Bengal, India.

Systomus clavatus: Pethiyagoda et al., 2012: 77 (valid).

Material examined. RGUMF-unregistered, 1 ex., 51.5 mm SL, Kameng River at Tippi, East Kameng district, Arunachal Pradesh (specimen damaged).

Vernacular names in Arunachal Pradesh. *Ngogi-Getak* (in Nyishi & Galo), *Taga* (in Adi).

Diagnosis. Mouth sub-inferior, cleft of mouth extending close to the level of anterior margin of the orbit. Barbels two pairs: one each of maxillary and rostral, snout long with many tubercles on tip. Dorsal fin with three simple and eight branched rays, last simple ray of dorsal fin osseous and serrated posteriorly, and its height slightly more than body depth at dorsal origin. Predorsal scales 15–16, lateral line scales 40–42, and transverse scales between the dorsal fin origin and lateral line 8.

Coloration. Bluish superiorly becoming white beneath, fins pale bluish white.

Distribution. Known from the Ganga-Brahmaputra basin in India: Assam, Arunachal Pradesh, West Bengal, Sikkim, and Barak basin in Manipur; Nepal and Bangladesh.

Remarks. This species was earlier under the genus *Poropuntius* (Vishwanath et al., 2007).

Economic importance. A popular food fish.

Conservational status: A widely spread species occurring in the Ganga-Brahmaputra and Barak River basins. IUCN status: Not assessed (NA).

31. Genus: *Oreichthys* Smith, 1933

Oreichthys Smith 1933: 63. Type species: *Oreichthys parvus* Smith 1933. Type by monotypy.

Body deep and stout, strongly compressed laterally; snout obtuse, no barbels; snout, interorbital region, cheeks, and opercle with rows of minute pores; abdomen anterior to pelvic-fin origin rounded, scales on lateral line larger than others; no gill rakers; eyes large, dorsolaterally placed, not visible from ventral surface, dorsal fin inserted above distal half of pectoral fin, no spinous ray in dorsal fin; complete or incomplete lateral line, caudal fin forked.

Oreichthys crenuchoides Schäfer 2009

Fig. 74.

FIG. 74 *Oreichthys crenuchoides*, RGUMF-334, 40.6 mm SL: lateral view.

Oreichthys crenuchoides, Schäfer 2009: 202, Figs. 1–2, 6–10; type locality: River Jorai, a tributary of Brahmaputra River, near Assam border, 26°27′35″N, 89°46′24″E, Coochbehar district, West Bengal, India.

Oreichthys crenuchoides: Knight, 2014: 5357 (valid).

Materials examined. RGUMF-332–333, 2 exs., 21.1–23.0 mm SL, Lakhimpur, Lakhimpur district, Assam. RGUMF-334, 1 ex., 40.6, Khari Nalah, Pakke Wildlife Sanctuary, East Kameng district, Arunachal Pradesh (AP). RGUMF-336, 1 ex., 40.6 mm SL, Doigurung River, Pakke Wildlife Sanctuary, East Kameng district, AP. RGUMF-unregistered, 1 ex. Dikrong River at Doimukh, Papum Pare district, AP.

Vernacular names in Arunachal Pradesh. *Ngogi-Getak* (in Nyishi & Galo), *Gaalic-metakk* (in Adi).

Diagnosis. No spot or blotch on anal fin, scales between the pelvic-fin origin and dorsal midline ½7½, snout blunt, a black blotch at caudal-fin base, lateral line incomplete, 2–5 perforating scales, 17–20 + 2 scales in longitudinal series, and 7–9 predorsal scales.

Coloration. In 10% formalin, body brownish and darker on back, a district blotch at the base of caudal fin present. Pectoral, pelvic, anal and caudal fin hyaline. Sub distal portion of second simple to second branched rays of dorsal fin blackish. In mature males dorsal fin is variegated with numerous grayish markings with a blackish streak at distal anterior margin.

Distribution. Known from the Brahmaputra River drainage in northern West Bengal, Assam, and Arunachal Pradesh.

Remarks. This study records *O. crenuchoides* for the first time from Assam and Arunachal Pradesh. We have observed some differences with the present examples and the original description: predorsal scale 7 ($n=4$) or 9 ($n=1$) (vs. 7 in the original description). Reports of *Oreichthys cosuatis* from Arunachal Pradesh by Nath and Dey (2000; 38, Fig. 31) is probably a misidentification of *O. crenuchoides* because the figure shows a distinct spot at the base of caudal fin (such spot is absent in *O. cosuatis*).

Economic importance. A highly popular ornamental fishes.

Conservational status: A widely spread species occurring in the Brahmaputra basin only. IUCN status: Data deficient (DD).

32. Genus: *Puntius* Hamilton, 1822

Puntius (subgenus of *Cyprinus*) Hamilton 1822: 310, 388. Type species: Masc. *Cyprinus sophore* Hamilton 1822. Type by subsequent designation.

Adult size small, not greater than 120mm SL; maxillary barbel may be absent or present, no rostral barbel; dorsal fin with iii,8 rays; anal fin with iii,5 rays; last simple dorsal-fin ray weak to moderately strong, unserrated; complete lateral line, usually with 22–28 pored scales on flank; free uroneural attached ventrolaterally to distal tip of pleurostyle; gill rakers simple, a post-epiphyseal fontanelle present; supraneurals 4; infraorbital 3 slender; 5th ceratobranchial narrow; 5+3+2 pharyngeal teeth; 12–14 abdominal and 14–16 caudal vertebrae; and body color pattern usually with a blackish spot or blotch on the caudal peduncle.

Key to the species of *Puntius*:

1a. Barbels absent, 2 black blotches: one each at the mid-base of dorsal fin and caudal peduncle .... *P. sophore*

1b. One pair of barbel present, 3 black blotches: one each at the dorsal-fin origin, caudal peduncle, and behind the opercle.. *P. chola*

***Puntius sophore* (Hamilton 1822)**
Fig. 75.

FIG. 75 *Puntius sophore*, RGUMF-48, 47.0mm SL: lateral view.

Cyprinus sophore, Hamilton 1822: 310, 389, pl. 19, Fig. 86; type locality: Srimangal [=Sreemangal] town market, from Hail Hoar floodplain near Moulvibazaar, Bangladesh [original type locality: Ganges River basin, India].

Puntius sophore, Kottelat, 2013: 149 (valid).

Material examined. RGUMF-48, 15 exs., 45.0–48.0 mm SL, Dikrong River at Doimukh, Papum Pare district, Arunachal Pradesh.

Vernacular names in Arunachal Pradesh. *Ngogi-Getak* (in Nyishi & Galo), *Gaalic-metakk* (in Adi).

Diagnosis. Body stout and laterally compressed. Mouth subterminal. Barbels absent. Dorsal fin inserted slightly behind pelvic-fin origin, with four unbranched rays and eight branched rays, last simple ray smooth and strong. Lateral line complete with 23–24+1 pored scales, transverse scale row 4½/1/3½, predorsal scales 8–10, circumpeduncular scales 12. A humeral mark and another caudal blotch present.

Coloration. In live, body silvery, with a blackish blotch on the caudal peduncle and another behind the opercle. Dorsal fin with a small black blotch at the base of 5–7 branched rays.

Distribution. Widely distributed throughout India, Nepal, Bangladesh, Pakistan, Myanmar, Thailand, and Yunnan (China). In Arunachal Pradesh: Ranga, Dikrong, Sinki (a tributary of the Siang River), and Siren rivers.

Economic importance. A highly popular ornamental fishes.

Conservational status: This fish is a widespread species occurring in multiple river basins. The fish is found in all lowland slow moving rivers and stagnant water bodies and breeds naturally even in small ponds and streams. IUCN status: Least concern (LC).

***Puntius chola* (Hamilton 1822)**

Fig. 76.

FIG. 76 *Puntius chola*, RGUMF-46, 55.6 mm SL: lateral view.

Cyprinus chola Hamilton, 1822: 312, 389; type locality: India: Northeastern part of Bengal, India. *Puntius chola*: Kottelat, 2013: 149 (valid).

Material examined. RGUMF-46, 18 exs., 45.0–76.0 mm SL, Rang River at Kimin, Papum Pare district, Arunachal Pradesh.

Vernacular names in Arunachal Pradesh. *Ngogi-Getak* (in Nyishi & Galo), *Gaalic-metakk* (in Adi).

Diagnosis. Body fairly deep and compressed. Humeral mark and caudal blotch prominent. Lateral line complete with 24–28 scales, transverse scale row 5½/1/3½, predorsal scales 11, pre-pelvic scales 11, preanal scales 14. Dorsal fin with iii,9 rays, last simple ray smooth and strong.

Coloration. Body silvery with olive green dorsally, caudal blotch present between 21st and 23rd scales. A black spot may or may not be present behind operculum. During breeding season, body marked with a broad crimson stripe across snout tip to end of caudal fin.

Distribution. Widely distributed throughout India, Nepal, Bangladesh, Sri Lanka, Myanmar, and Pakistan. In Arunachal Pradesh: Ranga River, Dikrong River and its tributary Senkhi River, etc.

Economic importance. A highly popular and demanded ornamental fish.

Conservational status: This fish is a widely spread species occurring in multiple river basins. The fish is found in all lowland slow moving rivers and stagnant water bodies and breeds naturally even in small ponds and streams. IUCN status: Least concern (LC).

33. Genus: *Pethia* Pethiyagoda et al., 2012

Pethia Pethiyagoda, Meegaskumbura & Maduwage 2012: 80. Type species: Fem. *Barbus nigrofasciatus* Günther 1868. Type by original designation.

Small adult size usually less than 80 mm SL, third infraorbital broad, maxillary barbels minute or absent, rostral barbels absent, last unbranched dorsal-fin ray stiff and serrated, lateral line usually incomplete, body color pattern usually consisting of a black blotch or spot anteriorly on the flank (absent in *Pethia conchonius*) and a black blotch on the caudal peduncle; and lacking post-epiphyseal fontanelle (Pethiyagoda et al., 2012).

Last unbranched dorsal-fin ray serrated; infraorbital three deep; rostral barbels absent.

Key to the species of *Pethia*:

1a. Lateral line complete, presence of blackish spot at dorsal and anal fin origin.......... *P. guganio*

1b. Lateral line incomplete, no blackish spots at dorsal and anal-fin origin................2

2a. Presence of a black band at the middle of caudal peduncle, 2½ scales between lateral-line row and pelvic-fin origin .... *P. gelius*

2b. No black band at the middle of caudal peduncle, 4–4½ scales between lateral-line row and pelvic-fin origin 3

3a. Presence of humeral spot.. *P. ticto*

3b. Absence of humeral.. ......*P. conchonius*

***Pethia conchonius* (Hamilton 1822)**
Fig. 77.

FIG. 77 *Pethia conchonius*, RGUMF-195, 44.5 mm SL: lateral view.

Cyprinus conchonius Hamilton, 1822: 317, 318, 389; type locality: ponds of northeast Bengal, and Kosi and Ami rivers, India.

Pethia conchonius: Lalramliana et al., 2014: 372 (valid).

Materials examined. RGUMF-47, 25 exs., 33.0–50.0 mm SL Poma River at Poma, Papum Pare district, Arunachal Pradesh (AP). RGUMF-195, 7 exs., 44.5–62.5 mm SL, Singen River, East Siang district, AP. RGUMF-unregistered, 30 exs., 38.5–48.0 mm SL, Ranga and Dikrong rivers, Papum Pare district, AP, Siang River at Pasighat, East Siang district, AP.

Vernacular names in Arunachal Pradesh. *Ngogi-Getak* (in Nyishi & Galo), *Gaalic-metakk* (in Adi).

Diagnosis. Humeral spot and barbels absent; last simple ray of dorsal fin serrated, incomplete lateral line, 5½ scales between the dorsal-fin origin and lateral-line row, total vertebrae 29.

Coloration. Body silvery yellowish with a blackish round midlateral blotch above the posterior end of anal-fin base, pelvic and anal fin proximally orange with blackish distal margin. Dorsal fin hyaline with blackish streak on anterodistal half of fin.

Distribution. Known widely in India from the Ganga-Brahmaputra, Mahanadi, Godavari, Krishna, and Cauvery river systems and other west-flowing rivers; Nepal, Bangladesh, Pakistan, and Afghanistan. In Arunachal Pradesh: see materials examined.

Economic importance. A highly popular and demanded ornamental fish.

Conservational status: This fish is a widely spread species occurring in multiple river basins. The fish is found in all lowland slow moving rivers and stagnant water bodies and breeds naturally even in small ponds and streams. IUCN status: Least concern (LC).

***Pethia ticto* (Hamilton 1822)**
Fig. 78.

FIG. 78 *Pethia ticto*, RGUMF-49, 42.0 mm SL: lateral view.

Cyprinus ticto Hamilton, 1822, 314, 389, pl. 8, Fig. 87; type locality: south-eastern parts of West Bengal, India.

Pethia ticto: Pethiyagoda et al., 2012: 81 (valid).

Materials examined. RGUMF-49, 11 exs., 35–47 mm SL, Poma River at Poma, Papum Pare district, Arunachal Pradesh. RGUMF-unregistered, 01 ex., 42.0 mm SL, Dikrong River at Doimukh, Papum Pare district, Arunachal Pradesh.

Vernacular names in Arunachal Pradesh. *Ngogi-Getak* (in Nyishi & Galo); *Gaalic-metakk* (in Adi).

Diagnosis. Body deep, and laterally compressed; dorsal fin with iii.8 rays, last simple ray osseously long and serrated posteriorly with 15–18 serrae; lateral line incomplete with 6–11 pored scales, 23–26 scales in lateral-scale row; 9 pre-dorsal scales, and transverse scales rows on body with ½5/1/4½; two black spots on flank along the lateral line scales series: one covering the third and fourth lateral line scales and the second on the caudal peduncle within the 16th–19th scales of the lateral line scale row.

Coloration. Head and body sides silvery, dorsally grayish. Two rows of black spots scattered on the dorsal fin (additional body marking see above).

Distribution. Known from the Ganga-Brahmaputra River system in India, Nepal, and Bangladesh; also known from Pakistan and Sri Lanka. In Arunachal Pradesh: Poma, Ranga, Subansiri, Dikrong, Siang, Dibang, Lohit, Noadhing, and Buridhing rivers.

Economic importance. A highly popular and demanded ornamental fish.

Conservational status: This fish is a widely spread species occurring in multiple river basins. The fish is found in all lowland slow moving rivers and stagnant water bodies and breeds naturally even in small ponds and streams. IUCN status: Least concern (LC).

***Pethia guganio* (Hamilton 1822)**

Fig. 79.

FIG. 79 *Pethia guganio*, RGUMF-335, 36.9 mm SL: lateral view.

Cyprinus guganio Hamilton, 1822: 338, 392; type locality: Brahmaputra and Yamuna rivers, India.

Pethia guganio: Knight, 2013: 181 (valid).

Material examined. RGUMF-335, 01 ex., 36.9 mm SL, Khari Nalah, Pakke Wildlife Sanctuary, East Kameng district, Arunachal Pradesh.

Vernacular names in Arunachal Pradesh. *Ngogi-Getak* (in Nyishi & Galo), *Gaalic-metakk* (in Adi).

Diagnosis. Lateral line complete with 28–29 + 1 pored scales, transverse scales rows on body with ½5/1/4. Body with three distinct black spots: one at dorsal fin origin, another above the origin of anal-fin base, and the other at the base of caudal fin.

Coloration. Body silvery, dorsally darker. For additional body color features see above.

Distribution. Occurring in the Ganga-Brahmaputra basin in India and Bangladesh. In Arunachal Pradesh: see material examined.

Remarks. This species rarely encountered in Arunachal Pradesh.

Economic importance. A highly popular and demanded ornamental fish.

Conservational status: This fish is a widely spread species occurring in the Ganga-Brahmaputra river basins. The fish is found in all lowland slow moving rivers and stagnant water bodies and breeds naturally even in small ponds and streams. IUCN status: Least concern (LC).

***Pethia gelius* (Hamilton 1822)**

Fig. 80.

FIG. 80 *Pethia gelius*, RGUMF-343, 33.6 mm SL: lateral view.

Cyprinus gelius, Hamilton 1822: 320, 390; type locality: River Torsa at Totapara in Jalpaiguri, West Bengal, India [original type locality: the ponds and ditches of the northeast parts of West Bengal].

Pethia gelius: Knight, 2013: 173 (valid).

Material examined. RGUMF-342 to 343, 3 exs., 30.2–33.6 mm SL, Tengapani River at Namsai, Namsai district, Arunachal Pradesh.

Vernacular names in Arunachal Pradesh. *Ngogi-Getak* (in Nyishi & Galo), *Gaalic-metakk* (in Adi).

Diagnosis. Barbel absent, dorsal fin with 3 simple and 8½ rays, its third simple ray osseous and serrated posteriorly with 20–25 serrae. Lateral line incomplete, with 3–4 pored scales; 22–23 scales in lateral series, 8 predorsal scales; and body transverse scales rows with ½4/1/2½ (additional diagnostic body color features see below).

Coloration. Body reddish brown with silvery band along the side, a black band at the middle of caudal peduncle covering scales 19–21 in the longitudinal series, flank with three less distinct blackish blotches: one behind opercle, second below the dorsal fin, third above the anal fin. Origin and subsequent basal parts of dorsal, anal, and pelvic fins blackish, distally orange.

Distribution. Known from Ganga-Brahmaputra basin in India and Bangladesh. This species is also reported from Pakistan. In Arunachal Pradesh: abundantly found in the foothills areas (see material examined).

Remarks. Knight (2013) designated a neotype of this species to a specimens collected from the Torsa River at Totapara in Jalpaiguri, West Bengal, India.

Economic importance. Highly popular and demanded ornamental fish because of its small size and peculiar body colorations.

Conservational status: This fish is a widely spread species occurring in the Ganga-Brahmaputra river basins. The fish is found in all lowland slow moving rivers and stagnant water bodies and breeds naturally even in small ponds and streams. IUCN status: Least concern (LC).

34. Genus: *Schizothorax* Heckel, 1838

Schizothorax Heckel 1838: 11. Type species: *Schizothorax esocinus* Heckel 1838. Type by subsequent designation.

Body elongated and subcylindrical; rounded abdomen; head short, somewhat cone shaped and blunt; mouth inferior, wide and slightly arched; lips fleshy and continuous; margin of lower lip with free posterior edge forming a sucker; lower jaw with horny cartilage; barbels two pairs: one each of rostral and maxillary; dorsal fin inserted opposite to pelvic fins, its last simple ray osseous, strong, and serrated posteriorly, its height shorter than its lateral head length; scales very small, lateral line complete with up to 110 scales; anal fin in a sheath covered by an enlarged row of scales; caudal fin fork.

Key to the species of *Schizothorax*:

1a. Mouth transverse and slightly arched backward, lower labial fold interrupted and short.......... *S. richardsonii*

1b. Mouth deeply arched backward; lower labial fold uninterrupted and longer.............................. *S. progastus*

***Schizothorax richardsonii* (Gray 1832)**
Fig. 81.

FIG. 81 *Schizothorax richardsonii*, RGUMF-64, 142 mm SL: (A) lateral view, and (B) mouth part.

Cyprinus richardsonii Gray, 1832: 94; type locality: not known.
Schizothorax richardsonii: Gurung et al., 2013: 4883 (valid).

Materials examined. RGUMF-61–64, 15 exs., 30.0–160 mm SL, Sangti River in West Kameng district, Siren R. in East Siang district, Kele and Panye rivers in Papum Pare district, Arunachal Pradesh (AP). RGUMF- 367, 20 exs., 137.8–210.5 mm SL, Tenga River, West Kameng district, AP.

Vernacular names in Arunachal Pradesh. *Riibu* (in Apatani); *Ngokha* (in Nyishi), *Kada* (in Galo), *Ngoying* (in Adi).

Diagnosis. Snout long (length 37.8%–40.0% HL), sparsely tuberculated in adult males, smooth in females and juveniles; mouth: inferior, transverse, and slightly arched, width 38.7%–41.3% HL, lower jaw heavily cornified with a keratinized cutting edge and lower labial fold uninterrupted and short; predorsal length 47.3%–50.1% SL; dorsal fin with iii,8 rays, last simple ray long, serrated posteriorly with 20–22 serrae, head length 22.5%–22.8% SL, pectoral-fin length 85.0%–86.9% HL, body depth at dorsal origin 19.0%–22.2% SL; pectoral i,17 rays; pelvic i,11 rays; anal iii,5½ rays, gill rakers on the first arch 3+19=22 (n=1); a series of 14 (on each side) enlarged scales along the posterior abdominal midline and anal-fin base, and 86–110 scales on lateral line.

Coloration. In 10% formalin, body dark brown above flank and belly light brownish.

Distribution. Known from the Ganga-Brahmaputra basin in India (Assam, Arunachal Pradesh, Sikkim, West Bengal, and Uttarakhand); Bhutan; Nepal; and the Indus basin in Pakistan and Afghanistan. In Arunachal Pradesh: see materials examined section.

Remarks. This fish is the most abundantly available species among *Schizothorax* in Brahmaputra drainage of Arunachal Himalaya.

Economic importance. This species is a game fish of high altitude rivers.

Conservational status: This fish is a widely spread species occurring in the Ganga-Brahmaputra river basins. Reports from other neighboring basins need confirmation. IUCN status: Vulnerable (VU).

***Schizothorax progastus* (McClelland 1839)**

Fig. 82.

FIG. 82 *Schizothorax progastus*, RGUMF-382, 68.8 mm SL: (A) lateral view, and (B) mouth part.

Oreinus progastus McClelland, 1839: 274, 343, pl. 40, Fig. 4; type locality: Mountain streams at Simla, Upper Assam, elevation 5000–6000 ft, India.

Schizothorax progastus: Gurung et al., 2013: 4483 (valid).

Materials examined. RGUMF-382, 1 ex., 68.8 mm SL, Sangti River in West Kameng district, Arunachal Pradesh. RGUMF-383, 1 ex., 112.6 mm SL, Dibang River at Roing, Lower Dibang valley district, Arunachal Pradesh.

Vernacular names in Arunachal Pradesh. *Ngoliang* (in Apatani); *Ngoreng* (in Nyishi), *Ngoyii* (in Galo), *Kadang/Ngoying* (in Adi). RGUMF: unregistered, 1 ex., 114 mm SL, Siring River, Arunachal Pradesh, altitude 500 msl.

Diagnosis. Body elongated and abdomen rounded. Snout pointed; mouth inferior and deeply curved, lip thick and fleshy, lower labial fold uninterrupted and moderately longer. Barbels two pairs: one each of maxillary and rostral. A series of nine (on each side) enlarged scales along the posterior abdominal midline and anal-fin base. Dorsal fin with nine branched rays, its last simple ray spinous and posteriorly serrated (11serrae in a specimen of length 68.8 mm SL), anal: iii,6 , ventral: i,9–10 rays, pectoral: i,15 and caudal: i,9 + 8,i or i, 8 + 8,ii rays.

Coloration. Body silvery, light grayish above head, dorsal, anal and caudal fin light gray, pectoral and pelvic light orange.

Remarks. *Schizothorax progastus* differs from *S. richardsonii* in having a deeply curve mouth (vs. transverse or slightly curved), lower labial fold moderately longer (vs. shorter), and fewer (11 vs. 20–22) posterior serrations of dorsal spine.

Distribution. Known from the Ganga-Brahmaputra river system of India: Assam, Arunachal Pradesh, Uttarakhand, and Himachal Pradesh; Bhutan and Nepal. In Arunachal Pradesh: see materials examined section.

Economic importance. This species is a valuable game fish and relished as an important delicious food fish.

Conservational status: This fish is a widely spread species occurring in the Ganga-Brahmaputra river basins. IUCN status: Least concern (LC).

35. Genus: *Semiplotus* Bleeker, 1860

Semiplotus Bleeker 1860: 424. Type species: *Cyprinus semiplotus* McClelland 1839. Type by subsequent monotypy (also by absolute tautonymy).

Body deep, short, and compressed; rounded abdomen; snout thick, prominent; mouth wide, transverse, and inferior; presence of a row of six open pores on each sides passing across the snout toward orbit; one pair of maxillary barbels; the lower jaw has knob at the symphysis and covered with a thin cartilaginous sheath; dorsal fin long with 20–23 branched rays, last simple dorsal-fin ray osseous, may or may not be serrated posteriorly; lateral line complete with 32–36 scales.

***Semiplotus semiplotus* (McClelland 1839)**
Fig. 83.

FIG. 83 *Semiplotus semiplotus*, RGUMF-58, 140mm SL: lateral view.

Cyprinus semiplotus, McClelland 1839: 274, 346, pl. 37, Fig. 2; type locality: Brahmaputra River rapids, Upper Assam, India.

Semiplotus semiplotus: Vishwanath & Kosygin, 2000: 92–102 (valid).

Materials examined. RGUMF-58, 4 exs., 50.0–165.0mm SL; Dikrong River, Papum Pare district, Arunachal Pradesh (AP), India, March 3, 2005. RGUMF-304, 5 exs., 136.7–156.5mm SL; Sinkin River (Brahmaputra basin), Lower Dibang Valley district, AP., 27°54′026″N and 95°13′55″E, elevation 175m asl., February 24, 2016, collectors Rashmi Dutta and Party. RGUMF-305, 3 exs., 137.2–167.2mm SL; Sipu River (Brahmaputra basin), West Siang district, AP., India, 28°06′28″N and 94°79′59″E, April 5, 2013, coll. A. Darshan and Party.

Vernacular names in Arunachal Pradesh. *Ngogir-pech/Ngogi-getak* (in Nyishi), *Orpe* (in Adi), *Orse* (in Galo).

Diagnosis. Dorsal fin with four osseous simple and 23–25 branched rays, last simple ray of dorsal fin posteriorly smooth; a transverse row of 10–12 open pores across the snout, lateral line with 32–33 scales, 6 transverse scales rows between the lateral line and the dorsal-fin origin, 4 rows of scales below lateral line to pelvic-fin origin, 11–12 predorsal scales, 11 + 11 circumferential scales, 5 + 5 circumpeduncler scales, and 8 postanal scales. Lateral line complete.

Coloration. In live, body laterally silvery and blackish dorsally. A prominent grayish bar reflects below dorsal fin when the fish is in water (a swarm of this species recognized by seeing its grayish bar). This bar-shaped grayish reflection turned indistinct when removed from water. Pectoral, pelvic, and anal fins orange. The degree of deepness of this orange color differs seasonally and with age of the fish. Fin color is more prominent in smaller fishes. In 70% alcohol, dull silvery laterally and dorsally grayish. In 10% formalin, overall body appears brownish.

Distribution. India: Ganga-Brahmaputra basin in Arunachal Pradesh, Assam, Sikkim, and West Bengal; Nepal (Shrestha, 2008) and Bhutan (Gurung et al., 2013). The report of this species from Myanmar (Day, 1878) is unlikely.

Remarks. Bleeker (1860) established the genus *Semiplotus* to accommodate *Cyprinus semiplotus*. Howes (1982) synonymized *Semiplotus* under the genus

Cyprinion Heckel 1843 based on the similarity in the osteological structure of the jaw, shape, and size of the mouth. However, Bănărescu and Herzig (1995) recognized both *Cyprinion* and *Semiplotus* as valid genera and reassigned *C. semiplotus* to the later genus.

Economic importance. A highly prized food fish.

Conservational status: This fish is a widely spread species occurring in the Ganga-Brahmaputra river basins. IUCN status: Not assessed (NA).

36. Genus: *Tor* Gray, 1834

Tor Gray 1834: pl. 96 (v. 2). Type species: *Tor hamiltonii* Gray 1834 (= *Cyprinus tor* Hamilton 1822). Type by monotypy.

Body elongated and moderately compressed; tapers smoothly toward caudal peduncle; rounded abdomen; mouth inferior; upper jaw slightly protractile; lips fleshy continuous at the angle of mouth; lower lip with or without a median fleshy lobe below symphysis and the post-labial groove continuous; dorsal with 12–13 rays (8–9 branched rays), its origin located opposite to pelvic fin; last simple ray osseous and smooth; scales large; lateral line complete with scales fewer than 30; caudal fin deeply forked.

Key to the species of *Tor*:

1a. Length of head considerably greater than body depth at dorsal-fin origin ....*T. putitora*

1b. Head length (26.8%–28.8% SL) shorter than body depth (26.9%–30.6% SL) at dorsal-fin origin..........*T. tor*

***Tor putitora* (Hamilton 1822)**
Fig. 84.

FIG. 84 *Tor putitora*, RGUMF-35, 225 mm SL: lateral view.

Cyprinus putitora Hamilton, 1822: 303, 388; type locality: eastern parts of West Bengal, India.

Tor putitora: Arunkumar & Basudha, 2003: 274 (valid).

Material examined. RGUMF-34–35, 20 exs., 60.0–225 mm SL, Poma River at Ramgath, Papum Pare district, Arunachal Pradesh. Sipu River at Aalo, West Siang district, Arunachal Pradesh.

Vernacular names in Arunachal Pradesh. *Ngogi/Ngauch/Ngopi* (in Nyishi), *Ngadi* (in Nocte), *Rulbung* (in Adi), *Orb-tako* (in Galo).

Diagnosis. Body elongated and laterally compressed. Head long. Length of head considerably greater than body depth at dorsal fin origin: head length 29.5%–32.4% SL and body depth 12.1%–15.6% SL; 29–30 lateral line scales; lower arm of first gill arch with 12–13 rakers.

Coloration. Body dark greenish dorsally, yellowish golden flank, silvery white ventrally. Fins hyaline except distal margin of caudal fin orange.

Distribution. Known from India, Bangladesh, Bhutan, Nepal, Myanmar, and Pakistan. In Arunachal Pradesh: Poma, Ranga, Kalpangi, Kameng, Subansiri, Dikrong, Siang, Dibang, Lohit, Noadhing, Buridhing, etc., rivers (Nath and Dey, 2000).

Remarks. *Tor putitora* is the state fish of Arunachal Pradesh.

Economic importance. Game or sport fish of regional importance.

Conservational status: This fish is a widely spread species (see distribution section). The fish is also occasionally available in some local fish markets in the states of Arunachal Pradesh and Assam. IUCN status: Endangered (EN).

***Tor tor* (Hamilton 1822)**

Fig. 85.

FIG. 85 *Tor tor*, RGUMF-33, 140.0mm SL: lateral view.

Cyprinus tor Hamilton, 1822: 338; type locality: Mahananda River, West Bengal, India.

Tor tor: Kottelat, 2013: 169 (valid).

Material examined. RGUMF-30–33, 22 exs., 52.0–160mm SL, Poma River at Poma, Ranga River at Yazali, Kalpangi, Kameng at Balukpung, Arunachal Pradesh.

Vernacular names in Arunachal Pradesh. *Ngorik* (in Nyishi), *Ngarat* (in Nocte), *Rulbung* (in Adi); *Gase* (in Galo).

Diagnosis. Body elongated and fairly deep. Head length (26.8%–28.8% SL) equal to or less than body depth (26.9%–30.6% SL). Mouth inferior, lips fleshy with uninterrupted labial fold. Barbels two pairs, maxillary barbel longer than the eye diameter, rostral barbel shorter. Dorsal fin with four simple and nine branched rays, last simple ray spinous and smooth. Scales large, lateral line complete with 26 scales, transverse scale rows 4–4.5/1/2.5–3, circumferential scales 20, and circumpeduncular scales 12.

Coloration. In live, body dorsally grayish, flanks golden yellow, ventrally silvery. Fins yellowish.

Distribution. Widely distributed in India (Ganga-Brahmaputra basin in northeast to Uttarakhand, Barak basin in Manipur); Bangladesh; Bhutan; Nepal; Myanmar; and Pakistan. In Arunachal Pradesh, the species is encountered at several locations: Poma River (R.) at Poma; Ranga R. at Yazali; Kalpangi R. at Lower Subansiri district, and Kameng at Balukpung.

Economic importance. Game or sport fish of regional importance.

Conservational status: This fish is a widely spread species (see distribution section). The fish is occasionally sold at the road side by some local fisherman in the states of Arunachal Pradesh and Assam. IUCN status: Near threatened (NT).

Subfamily: Squaliobarbinae

37. Genus: *Ctenopharyngodon* Steindachner, 1866

Ctenopharyngodon Steindachner 1866: 782. Type species: *Ctenopharyngodon laticeps* Steindachner 1866. Type by monotypy.

Body elongated, subcylindrical, and compressed posteriorly; round abdomen; head flattened; mouth terminal; simple lips on jaws, upper jaw slightly extends over the lower jaw and protractile; barbels absent; eyes large; snout short and its length less than or equal to the eye diameter; dorsal fin with three simple and seven branched rays, no osseous simple rays, its origin located ahead of pelvic fin and close to snout-tip than caudal-fin base; scales large and cycloid, lateral line complete with 40–42 scales; caudal fin is deeply forked.

***Ctenopharyngodon idella* (Valenciennes 1844)**
Fig. 86.

FIG. 86 *Ctenopharyngodon idella*, RGUMF-unregistered, 280 mm SL: lateral view.

Leuciscus idella Valenciennes 1844: 362; type locality: China.

Ctenopharyngodon idella: Vishwanath et al., 2007: 41 (valid).

Material examined. RGUMF-unregistered. 2 exs., 75.3–280 mm SL, purchase from Doimukh Bazar, Papum Pare district, Arunachal Pradesh.

Vernacular names in Arunachal Pradesh. Ngogi/*Grass carp* (in Nyishi, Adi, & Galo).

Diagnosis. Body elongated, subcylindrical. Head depressed. Mouth terminal. Dorsal fin short and placed slightly anterior to pelvic fin insertion, dorsal spine weak and smooth. Lateral line complete and with 40–42 scales, maximum size reaching up to ca. 800 mm SL.

Coloration. Body silvery, grayish above head.

Distribution. China and Russia; widely introduced elsewhere including India. In Arunachal Pradesh: cultured in ponds.

Remarks. This fish is introduced species in Arunachal Pradesh.

Economic importance. This species is a food fish.

Conservational status: This species is widely introduced elsewhere. The fish is also occasionally available in some local fish markets in the states of Arunachal Pradesh. IUCN status: Not assessed (NA).

Subfamily: Xenocyprinae

38. Genus: *Hypophthalmichthys* Bleeker 1860

Hypophthalmichthys Bleeker 1860: 433. Type species: Masc. *Leuciscus molitrix* Valenciennes 1844. Type by subsequent designation.

Body robust and compressed; head large; eyes are located near the ventral side; mouth terminal, large, wide, cleft not extending to anterior rim of orbit; barbel absent; adipose fin absent; dorsal fin inserted behind pelvic fin, with 10 rays (7 branched). Anal fin with 12–14 branched; scales small and cycloid; lateral line decurved and complete with 110–115 scales, caudal fin forked.

Key to the species of *Hypophthalmichthys*:

1a. Abdominal edge keeled from throat to vent. .... *H. molitrix*

1b. Abdominal edge keeled from pelvic fin to vent (anus)............................ ....*H. nobilis*

***Hypophthalmichthys molitrix* (Valenciennes 1844)**
Fig. 87.

FIG. 87 *Hypophthalmichthys molitrix*, RGUMF-unregistered, 200 mm SL: lateral view.

Leuciscus molitrix Valenciennes 1844: 360, type locality: China.

Hypophthalmichthys molitrix: Vishwanath et al., 2007: 12 (valid).

Material examined. RGUMF-unregistered. 2 exs., 102.6–200 mm SL, purchase from Doimukh Bazar, Papum Pare district, Arunachal Pradesh.

Vernacular names in Arunachal Pradesh. *Silver carp* (in Nyishi, Adi, & Galo).

Diagnosis. Body elongated and compressed. Head very large. Mouth terminal. Dorsal fin inserted slightly posterior to pelvic origin. Abdominal edge keeled from throat to vent. Scales small with 113–116 scales on lateral line.

Coloration. Body silvery, grayish above.

Distribution. China and Russia; introduced elsewhere, including India. In Arunachal Pradesh: cultured in ponds.

Remarks. Introduced species.

Economic importance. This species is a food fish.

Conservational status: This species is widely introduced elsewhere. The fish is also occasionally available in some local fish markets in the state of Arunachal Pradesh. IUCN status: Near threatened (NT).

***Hypophthalmichthys nobilis* (Richardson 1845)**

Fig. 88.

FIG. 88 *Hypophthalmichthys nobilis*, RGUMF-unregistered, 290 mm SL: lateral view.

Leuciscus nobilis Richardson 1845: 140, pl. 63, Fig. 3; type locality: Canton, China.

Hypophthalmichthys nobilis: Vishwanath et al., 2007: 12 (valid).

Material examined. RGUMF-unregistered. 1 exs., 290 mm SL, purchase from Doimukh Bazar, Papum Pare district, Arunachal Pradesh.

Vernacular names in Arunachal Pradesh. *Silver carp* (in Nyishi, Adi, & Galo).

Diagnosis. Body elongated and compressed. Head very large. Mouth terminal. Dorsal fin inserted slightly posterior to pelvic origin. Abdominal edge keeled from pelvic fin to vent (anus). Scales small with 113–116 scales on lateral line.

Coloration. Body silvery, grayish above.

Distribution. China and Russia; introduced elsewhere, including India. In Arunachal Pradesh: cultured in ponds.

Remarks. Introduced species.

Economic importance. This species is a food fish.

Conservational status: This species is widely introduced elsewhere. The fish is also occasionally available in some local fish markets in the state of Arunachal Pradesh. IUCN status: Data deficient (DD).

5. Family: Psilorhynchidae

39. Genus: *Psilorhynchus* McClelland, 1838

Psilorhynchus McClelland 1838: 944. Type species: Masc. *Cyprinus sucatio* Hamilton 1822. Type by subsequent designation.

Small adult size, body cylindrical, and depressed anteriorly; head broadly pointed anteriorly; snout rounded; mouth small and inferior, bordered by a deep longitudinal groove on each side; rostral cap and upper lip fused; lower jaw covered by a thick cushion; chest naked, scales small to moderate, no barbels, lateral line with 34–47 scales; air bladder not enclosed in bony capsule.

Key to the species of *Psilorhynchus*:

1a. Abdomen fully scaled; body sides without blotches; snout long and spatulated .... *P. sucatio*

1b. Abdomen partially scaled or naked; body sides with blotches; snout short or moderate and rounded...2

2a. Lateral line scales 30–34.. ....*P. balitora*

2b. Lateral line scales 37–44.. 3

3a. Branched caudal fin rays 17–18 (9 + 8–9); total vertebrae 43–45 .... *P. arunachalensis*

3b. Branched caudal fin rays 15 (8–7), total vertebrae 41–42................*P. homaloptera*

***Psilorhynchus balitora* (Hamilton 1822)**

Fig. 89.

FIG. 89 *Psilorhynchus balitora*, RGUMF-82, 60.2 mm SL: (A) lateral view, and (B) mouth part.

Cyprinus balitora Hamilton, 1822: 348, 394; type locality: Northeastern Bengal, India.

Psilorhynchus balitora: Lalramliana et al., 2014: 270 (valid).

Material examined. RGUMF-82, 25 exs., 40.0–65.0 mm SL, Sipu River at Doji, East Siang district, Arunachal Pradesh.

Vernacular names in Arunachal Pradesh. *Jarzab* (in Nyishi), *Ngokapagra* (in Galo), *Daral* (in Adi).

Diagnosis. Dorsal fin with 8 branched rays; 30–34 scales along lateral line; 3/1/2 lateral transverse scales rows; simple pectoral fin 6–7; circumferential scales 18; air bladder naked with anterior and posterior chamber.

Coloration. Body grayish with irregular black bands.

Distribution. Known only from the Ganga-Brahmaputra drainages in India (Arunachal Pradesh, Assam, and West Bengal), Nepal, and Northeast Bangladesh.

Economic importance. This species has high ornamental value.

Conservational status: This fish is a widely spread species occurring in the Ganga-Brahmaputra river basins. IUCN status: Least concern (LC).

Psilorhynchus homaloptera **Hora and Mukerji 1935**

Fig. 90.

FIG. 90 *Psilorhynchus homaloptera*, RGUMF-84, 77.2 mm SL: (A) lateral, (B) dorsal views, and (C) mouth part.

Psilorhynchus homaloptera Hora & Mukerji, 1935: 391–394; type locality: Keleki stream, Emilomi, Nagaland.

Psilorhynchus homaloptera: Shangningam & Vishwanath, 2016 (valid).

Material examined. RGUMF-83–85, 18 exs., 42.0–80.0 mm SL, Siren River in East Siang district, Kalpangi River in Lower Subansiri district and Dikrong River in Papum Pare district, Arunachal Pradesh.

Vernacular names in Arunachal Pradesh. *Ngocher/Jarzab* (in Nyishi & Galo), *Daral* (in Adi).

Diagnosis. Body well depressed and flattened with depth 13.0%–17.0% SL at dorsal origin, ventral surface of body between pair fins scaleless except one scale anterior to anal opening; pectoral fin with 8 simple and 7–9 branched rays, predorsal scales 12–14, lateral line scales 37–40; transverse scale rows 3½/1/2½, scale rows around caudal peduncle 10, and principal caudal-fin rays 9 + 8.

Coloration. In freshly preserved specimens (in 10% formalin), dorsal and lateral surface of body brownish and creamish ventrally, 7–8 round to squarish dark brown lateral blotches arranged along flank. Occiput dark brown.

Distribution. India: Nagaland, Manipur, Meghalaya, and Arunachal Pradesh; Nepal, and Myanmar. In Arunachal Pradesh: Sireng (a tributary the Siang River in Siang, Upper Siang district); Kalpangi (a tributary of the Subansiri River); and Dikrong rivers.

Economic importance. This species has high ornamental value.

Conservational status: This fish is a widely spread species occurring in the Brahmaputra and Barak river basins. IUCN status: Least concern (LC).

***Psilorhynchus arunachalensis* (Nebeshwar, Bagra and Das 2007)**
Fig. 91.

FIG. 91 *Psilorhynchus arunachalensis* paratype RGUMF-02, 106 mm SL: (A) lateral view, and (B) mouth part.

Psilorhynchoides arunachalensis Nabeshwar, Bagra, & Das, 2007: 2632; type locality: Dirang River at Dirang West Kameng district-Brahmaputra river system.

Psilorhynchoides arunachalensis: Lalramliana et al., 2014: 270 (valid).

Materials examined. RGUMF-01, holotype, 92.0 mm SL, Male, West Kameng district, Dirang River at Dirang (Brahmaputra river system), Arunachal Pradesh, India, Kenjum Bagra August 24, 2005. RGUMF-02, paratypes, 3 exs., 104.0–106 mm SL, Dirang River at Dirang (Brahmaputra river system), West Kameng district, Arunachal Pradesh, India, Kenjum Bagra, August 24, 2005. RGUMF-03; paratypes, 4 exs., 65.7–77.2 mm SL, Dikrong River at Doimukh, Papum Pare district, Arunachal Pradesh, Gejom Lollen, 3 March 2005. RGUMF-04, paratypes, 7 exs., 65.3–70.0 mm SL, Siang River at Pasighat, East Siang district, Arunachal Pradesh, India, Karsen Nyori, February 7, 2005. RGUMF-05, paratypes, 3 exs., 60.0–76.0 mm SL, Sireng River at Rottung, East Siang district, Arunachal Pradesh, India, Kenjum Bagra, September 31, 2005.

Vernacular names in Arunachal Pradesh. *Ngocher/Jarzab* (in Nyishi), *Ngoke* (in Galo), *Daral* (in Adi).

Diagnosis. Pectoral fin with 8–9 simple and 8–10 branched rays. Pelvic fin moderately enlarged with 2 simple and 7 branched rays. Predorsal scales 13–18, Lateral line with 42–44 + 3–4 scales, transverse scale rows: 3½–4½/1/2½–3½, scales absent on chest, total vertebrae 43–45.

Coloration. In 10% formalin, head and body dorsally grayish, laterally yellowish-brown, and ventrally yellowish-white. A series of 6–12 dark blotches on flank. Pair fins dorsally grayish to yellowish-brown and ventrally yellowish-white.

Distribution. This species is known from various rivers of Arunachal Pradesh viz. Kameng, Dikrong, and Siang River systems (all belong to the Brahmaputra basin).

Remarks. In some type specimens of this species, some black spots are observed on the body and fins. These spots are in fact skin infections. We have encountered such infection is *Schistura* sp. and *Schizothorax* sp., which were collected from the Dirang and Tenga rivers (all tributaries of the Kameng river system).

Economic importance. This species has high ornamental value.

Conservational status: This species is widely distributed in Arunachal Pradesh state. IUCN status: Data deficient (DD).

Psilorhynchus sucatio **(Hamilton 1822)**

Fig. 92.

FIG. 92 *Psilorhynchus sucatio*, RGUMF-402, 66.2 mm SL: lateral view.

Cyprinus sucatio, Hamilton, 1822: 347, 393; type locality: Northern Bengal, India.

Psilorhynchus sucatio: Lalramliana et al., 2015: 177 (valid).

Material examined. RGUMF-402, 01 ex., 66.2 mm SL, Siang River at Pasighat, East Siang district, Arunachal Pradesh.

Vernacular names in Arunachal Pradesh. *Ngocher/Jarzab* (in Nyishi), *Ngoke* (in Galo), *Daral* (in Adi).

Diagnosis. Snout long and spatulated. Dorsal fin with 2 simple and 8 branched rays. Pectoral-fin rays 4–5 simple and 7–9 branched rays, and its length equal to pelvic-fin length. Body scales cycloid, moderately large, with 3–4 prominent radii over anterior and posterior fields of scale. Abdomen fully scaled. Lateral line scales 33–36; and post-epiphyseal fontanelle absent.

Coloration. In 10% formalin, body ground color brownish dorsally and lighter cream ventrally, 6–8 round to squarish dark brown lateral blotches arranged along flank, 8–9 dorsal saddles along the back of which the one across the origin of dorsal fin is darkest. Dorsal fin with 2–3 rows of blackish spots arranged horizontally.

Distribution. Known from northeastern India, Nepal and Bangladesh.

Remarks. This species is easily distinguish from other congeners by its spatulated snout.

Economic importance. This fish has high ornamental value.

Conservational status: This fish is a widely spread species occurring in the Ganga-Brahmaputra river Basin. IUCN status: Least concern (LC).

6. Family: Botiidae

40. Genus: *Botia* Gray, 1831

Botia Gray 1831: 8. Type species: *Botia almorhae* Gray 1831. Type by monotypy.

Body short, compressed, moderately deep, and rounded abdomen; head long and pointed, snout conical, mouth sub-inferior, small and narrow; eyes moderately large, anterior nostrils tubular; lips thick, and fleshy; a bifid erectile suborbital spine below or in front of eyes; barbels 6–8, 4 rostral which are joined at their bases; dorsal fin inserted above origin of pelvic fin or slightly ahead; lateral line present; no scales on head. Caudal fin forked or deeply emerginated.

Key to the species of *Botia*:

1a. Body sides with 7–8 blackish bands descending from the back toward the abdomen ....*B. dario*
1b. Body sides with irregular reticulation of brownish to black bands.........
.............................*B. rostrata*

***Botia dario* (Hamilton 1822)**
Fig. 93.

FIG. 93 *Botia dario*, RGUMF-96, 60.6 mm SL: lateral view.

Cobitis dario Hamilton, 1822: 354, 394, pl. 29, Fig. 95; type locality: Bengal.
Botia dario: Kottelat 2012: 15 (valid).

Material examined. RGUMF-96, 5 exs., 55.0–115 mm SL, Panye & Ranga rivers at Kimin, Papum Pare district, Arunachal Pradesh.

Vernacular names in Arunachal Pradesh. *Ngora-poga* (in Nyishi) *Pan-ngoi* (in Galo), *Ribii* (in Adi).

Diagnosis. Body short and compressed; a bifid suborbital spine below eye; dorsal fin with 12–13 rays and anal fin with 7–8 rays; barbels four pairs: two pairs of rostral, one pair each of maxillary and mandibular (see coloration for additional diagnostic features).

Coloration. Body ground color yellowish gold, flank with blackish bands descending from the back toward the abdomen: bands anterior to or below dorsal fin descent obliquely, and post dorsal bands descent vertically; head with two broad bands: one between orbit and another one between internarial to tip of snout; and two or three cross bands either lobe of the caudal fin.

Distribution. Known from the Ganga-Brahmaputra River drainages in India, Nepal, Bhutan, Bangladesh. In Arunachal Pradesh: Kameng, Subansiri, Siang, Dibang, Lohit, Tirap, Noadhing, Buridhing rivers (Nath and Dey, 2000). Additional data see material examined section.

Economic importance. This species has high ornamental value.

Conservational status: This fish is a widely spread species occurring in the Ganga-Brahmaputra river Basin. IUCN status: Least concern (LC).

***Botia rostrata* Gunther 1868**
Fig. 94.

FIG. 94 *Botia rostrata*, RGUMF 97, 98.7 mm SL: lateral view.

Botia rostrata, Günther 1868: 367; type locality: West Bengal, India.
Botia rostrata: Zhang et al., 2016: 154 (valid).

Materials examined. RGUMF-97–98, 17 exs., 60.0–115 mm SL, Panye River and Ranga River at Kimin, Papum Pare district, Arunachal Pradesh. RGUMF-unregistered specimen: 25 exs., 75.2–114.6 mm SL: Sinkin River at Anpum, Lower Dibang Valley district; Senkhi River (a tributary of the Dikrong River) at Itanagar; Dikrong River at Doimukh; all in Papum Pare district, Arunachal Pradesh.

Vernacular names in Arunachal Pradesh. *Pan-ngoi/Ngora-poga* (in Nyishi), *Rani mashi* (In Galo), *Ribii* (in Adi).

Diagnosis. Body ground color yellowish to golden, superimposed with irregular reticulation of brownish to black bands, head with a dark bar under the eye, one on the opercle and another across the snout. Fins yellowish to orange, dorsal fin crossed with three (basal, middle, and distal) blackish bands, caudal fin with 4–5 bands, and pectoral fin with 3–4 thin bands. Fin counts: dorsal fin with ii–iv,8–10 rays, pectoral fin with i,11–12 rays, pelvic fin with i,6–7 rays and anal fin with i–iii,5–7, caudal fin fork with 19 principal rays.

Coloration. In freshly preserved adult specimens (in 10% formalin), body base color yellowish to golden with greenish to bluish tinge, superimposed with irregular reticulation of brownish to black bands. On head, a dark bar under the eye, one on the opercle and another across the snout, a dark band across the interorbital encircle the eye, then two horizontal stripes: one from the above and another below the orbital rim runs toward the snout tip (for additional coloration features see diagnosis section).

Distribution. Known from the Ganga-Brahmaputra basin in Northeast India and West Bengal; Bangladesh; and China (Yunnan). In Arunachal Pradesh: Kameng, Subansiri, Siang, Dibang, Lohit, Tirap, Noadhing, and Buridhing rivers (Nath and Dey, 2000). Additional distribution data: see materials examined section.

Remarks. Records of this species from the Irrawaddy and Salween basins need confirmation.

Economic importance. This loach has high ornamental value.

Conservational status: This fish is a widely spread species occurring in the Ganga-Brahmaputra river Basin. IUCN status: Vulnerable (Vu).

7. Family: Cobitidae

41. Genus: *Canthophrys* Swainson, 1838

Canthophrys Swainson 1838: 364. Type species: *Canthophrys albescens* Swainson 1839 (=*Cobitis cucura* Hamilton 1822). Type by subsequent designation.

Body cylindrical, stout, elongate, and rounded abdomen; head swollen with a shallow oval depression on occiput; nostril tubular and closely placed, mouth small, inferior and horse shoe shaped; snout long, with soft warty tubercules; eyes large and bulging; a suborbital bifid spine present below eyes; lips fleshy, continuous, upper lip thickly papillated protruding over lower lip; no teeth on jaws and palate; barbels 3 pairs: 1 rostral and 2 maxillary; cycloid scales; lateral line complete.

***Canthophrys gongota* (Hamilton 1822)**

Fig. 95.

FIG. 95 *Canthophrys gongota*, RGUMF-99, 90.0mm SL: lateral view.

Cobitis gongota Hamilton, 1822: 351, 394; type locality: Northern West Bengal.

Cantophrys gongota: Kottelat 2012: 23 (valid).

Materials examined. RGUMF-99, 01 ex., 90mm SL, Dikrong at Doimukh, Papum Pare district, Arunachal Pradesh. RGUMF-unregistered, 3 exs., 79.0–88.9mm SL, Singen River at Saku-Kadu Village, Brahmaputra drainage, East Siang district of Arunachal Pradesh (AP).

Vernacular names in Arunachal Pradesh. Not available (in Nyishi & Galo), *Ribii* (in Adi).

Diagnosis. Body cylindrical and elongated, head lacks scales and wider than depth, body covered with large cycloid scales, eyes large and bulging with a bifid suborbital spine below, barbels three pairs: one pair rostral and two pairs maxillary, mouth small and inferior, dorsal fin inserted behind pelvic fins with 10 rays, lateral line complete, and caudal fin truncate.

Coloration. In live, body with six saddles along back: three predorsal, one across posterior end of dorsal fin base, and two post-dorsal, flank with six posterodorsally elongated midlateral blackish blotches, top of head with saddles between the eye and the occipital region, a dark blotches below the eye which spread on opercle, lateral blotches interconnected with saddles. Lower half of flank and ventral surface yellowish white. Dorsal and caudal fins with numerous transverse rows of blackish spots with orange background, other fins tinged orange without spots.

Distribution. *Canthophrys gongota* is known from the Ganga-Brahmaputra River basin in India (northeastern India to Uttar Pradesh), Bangladesh, and Nepal. In Arunachal Pradesh: see materials examined section.

Remarks. This fish is the only species known under the genus *Canthophrys*.

Economic importance. The fish has high ornamental value.

Conservational status: This fish is a widely spread species occurring in the Ganga-Brahmaputra river Basin. IUCN status: Least concern (LC).

42. Genus: *Lepidocephalichthys* Bleeker, 1863

Lepidocephalichthys Bleeker 1863: 38, 42. Type species: *Cobitis hasselti* Valenciennes 1846. Type by original designation.

Body elongated and moderately compressed; abdomen rounded; head short and conical; mouth inferior, lips thick, continuous at the angle of mouth; lower lip divided in the middle, no teeth on jaws and palate; eye lacks free orbital margin; a large bifid erectile suborbital spine present below or in front of eye; barbels three pairs: one each of rostral, maxillary, and mandibular; head scaled on cheek and operculum; dorsal fin inserted slightly in front of pelvic fin; lateral line absent; caudal fin truncated, emerginated, or forked.

Key to the species of *Lepidocephalichthys*:

1a. Caudal fin deeply forked.. *L. goalparensis*

1b. Caudal fin truncated or rounded or slightly emarginated or notched.. 2

2a. Caudal fin with two dark spots, one at the upper base of fin and one at the center of the distal border of fin.. *L. annandalei*

2b. Caudal fin with only one dark spot at upper base of fin 3

3a. Dorsal fin insertion directly above or just posterior to pelvic-fin origin; short predorsal length with 53%.. ...... *L. guntea*

3b. Dorsal fin inserted conspicuously behind to pelvic-fin origin; Long predorsal with 61% SL .... *L. arunachalensis*

Lepidocephalichthys guntea **(Hamilton 1822)**
Fig. 96.

FIG. 96 *Lepidocephalichthys guntea*, RGUMF-100, 68.3 mm SL: lateral view.

Cobitis guntea Hamilton, 1822: 353, 394; type locality: Bengal, India.
Lepidocephalichthys guntea: Havird & Page, 2010: 153 (valid).

Material examined. RGUMF-100, 17 exs., 43.0–72.0 mm SL Dikrong at Doimukh, Papum Pare district, Arunachal Pradesh.

Vernacular names in Arunachal Pradesh. *Remum-poda* (in Nyishi & Galo), *Ribii* (in Adi).

Diagnosis. Dorsal fin inserted just behind or directly above pelvic-fin origin. Lateral and ventral side of head naked. Caudal fin truncate with a dark spot just above the mid-base. Body with a midlateral row of dark spots in the form of band.

Coloration. In 10% formalin, pale white body background color in both sexes. In females, there are 8–14 dark spots or blotches along the side. In males, there is a single, very prominent dark stripe along the side. In some females, the spots may also be fused into a weak stripe. In both sexes, there are 3–7 predorsal and 3–6 postdorsal dark bars along the dorsum.

Distribution. Widely found in the Ganga-Brahmaputra River system in north India, Bangladesh, and Nepal. In Arunachal Pradesh: Kameng, Subansiri, Dibang, Noadhing, and Buridhing rivers (Nath and Dey, 2000), and also in the Dikrong and Siang rivers.

Economic importance. The fish has high ornamental value.

Conservational status: This fish is a widely spread species occurring in the Ganga-Brahmaputra river Basin. IUCN status: Least concern (LC).

Lepidocephalichthys annandalei **Chaudhuri 1912**
Fig. 97.

FIG. 97 *Lepidocephalichthys annandalei*, RGUMF-412, 47.4 mm SL: lateral view.

Lepidocephalichthys annandalei, Chaudhuri 1912: 442, pl. 40, Figs. 3, 3a–b; type locality: Mahananda River at Siliguri and Teesta River near Jalpaiguri, India.
Lepidocephalichthys annandalei: Havird & Page, 2010: 145 (valid).

Material examined. RGUMF-412, 5 exs., 39.1–47.4 mm SL, Dikrong River at Gumto, Papum Pare district, Arunachal Pradesh.

Vernacular names in Arunachal Pradesh. *Remum-poda* (in Nyishi & Galo), *Ribii* (in Adi).

Diagnosis. Dorsal fin inserted anterior to pelvic-fin origin. Scales are absent on the top and sides of head. Caudal fin: truncated or notched with two broad V-shaped bands, and bears two distinct black spots: one above the middle of the base of the caudal fin and another subdistally on the medial fin rays.

Coloration. In 10% formalin, body with a pale yellow background superimpose with dark patterns on dorsum and 6–12 spots along the side that may be fused into a stripe. The dorsal and anal fins are reticulated (additional color pattern see diagnosis).

Distribution. Known from the Ganga-Brahmaputra basin in India, Bangladesh, and Nepal. In Arunachal Pradesh: Kameng, Subansiri, Siang, Dibang, Noadhing, and Buridhing rivers (Nath and Dey, 2000), and also known from the Dikrong, Siang, and Subansiri rivers.

Remarks. Barbels shorter than its congeners: first rostral pair never extends to anterior nostril.

Economic importance. A small fish of high ornamental value.

Conservational status: This fish is a widely spread species occurring in the Ganga-Brahmaputra river basin. IUCN status: Least concern (LC).

Lepidocephalichthys arunachalensis Datta & Barman 1984

Fig. 98.

FIG. 98 *Lepidocephalichthys arunachalensis* RGUMF-387: (A) live colouration, 80.2 mm SL, and (B) preserved colouration, 81.1 mm SL.

Noemacheilus arunachalensis Datta & Barman 1984: 275, pl. 14; type locality: Namdapha River, Namdapha Wildlife Sanctuary, Tirap district, Arunachal Pradesh, India.

Lepidocephalichthys arunachalensis: Havird & Page, 2010: 154 (valid).

Materials examined. RGUMF-338, 5 exs., 61.8–77.7 mm SL, streams at Mebo, East Siang district, Arunachal Pradesh (AP). RGUMF-339, 5 exs.,

58.5–81.7 mm SL, Sinkin river at Anpum, Lower Dibang Valley district, AP. RGUMF-387, 05 exs., 75.1–81.1 mm SL, streams at Mebo, a tributary of Siang River, East Siang district, AP.

Vernacular names in Arunachal Pradesh. *Remum-poda* (in Nyishi & Galo), *Ribii* (in Adi).

Diagnosis. Dorsal fin inserted conspicuously posterior to pelvic-fin origin, caudal fin rounded or truncated with blackish reticulations, body sides with 11–14 large blotches, adult size reaching up to 81.7 mm SL.

Coloration. In live, head and body with light-orange ground color. Dorsum and head spotted with irregular grayish markings. Flanks with 11–14 blackish blotches. A dark distinct bar present at the base of caudal fin, which dorsally merged into a spot at the upper base of caudal fin. Fin rays of dorsal, pectoral fins, and caudal fin dotted with blackish spots, interradial membrane hyaline. Pelvic and anal fins hyaline.

Distribution. Known only from Arunachal Pradesh: Streams and rivers in East Siang district, Lower Dibang Valley district and Namdapha River in Tirap district.

Remarks. *L. arunachalensis* seems most similar to *L. guntea* and *L. berdmorei* but differs by its dorsal-fin origin being conspicuously located posterior to pelvic-fin origin.

Economic importance. This species has high ornamental value.

Conservational status: This fish is occurring in all foot hill streams and rivers of Arunachal Pradesh. IUCN status: Endangered (EN).

***Lepidocephalichthys goalparensis* Pillai and Yazdani, 1976**
Fig. 99.

FIG. 99 *Lepidocephalichthys goalparensis*, RGUMF-426, 54.6 mm SL: lateral view.

Lepidocephalichthys goalparensis, Pillai & Yazdani 1976: 15, Fig. 2; type locality: Under a culvert, about 6 km west of Dudhnai, Inspection Bungalow, Goalpara district, Assam State, India.

Lepidocephalichthys goalparensis: Kottelat, 2012: 32 (valid).

Material examined. RGUMF-426, 05 exs., 54.6–66.2 mm SL, Kameng River at Tippi, East Kameng district, Arunachal Pradesh.

Vernacular names in Arunachal Pradesh. *Remum-poda* (in Nyishi & Galo), *Ribii* (in Adi).

Diagnosis. Dorsal fin inserted at level through slightly anterior to pelvic-fin origin; a strongly forked caudal fin with 4–5 grayish broad V-shaped bars, lamina circularis of males formed by fusing the seventh and eighth pectoral rays. About 50 predorsal scales; no scales on top of head.

Coloration. In 10% formalin, body background creamy with blackish patterns. There are about 7–11 oval, dark spots on the flank, that seldom connect to form a line. Numerous irregular dark spots on the top of the head. A large, conspicuous dark spot present over the upper caudal-fin base, usually spans from the third to seventh rays. The caudal fin has 4–5 regularly spaced, dark, broad, V-shaped bars.

Distribution. Known from the Ganga-Brahmaputra river drainages in northern India, Nepal, Bangladesh, and Myanmar (Irrawaddy basin). In Arunachal Pradesh: occurring in Kameng drainage.

Economic importance. This species has high ornamental value.

Conservational status: This fish is a widely spread species occurring in the Ganga-Brahmaputra river basin. IUCN status: Least concern (LC).

43. Genus: *Pangio* Blyth, 1860

Pangio Blyth 1860: 169. Type species: *Cobitis cinnamomea* McClelland 1839 (=*Cobitis pangia* Hamilton 1822). Type by monotypy.

Body elongated, slender with uniform thickness; dorsal fin inserted nearly above anal fin with six branched rays; head very short, compressed and lacks scales; nostril tubular, vertebrae ranging between 44 and 72.

Pangio pangia (Hamilton 1822)

Fig. 100.

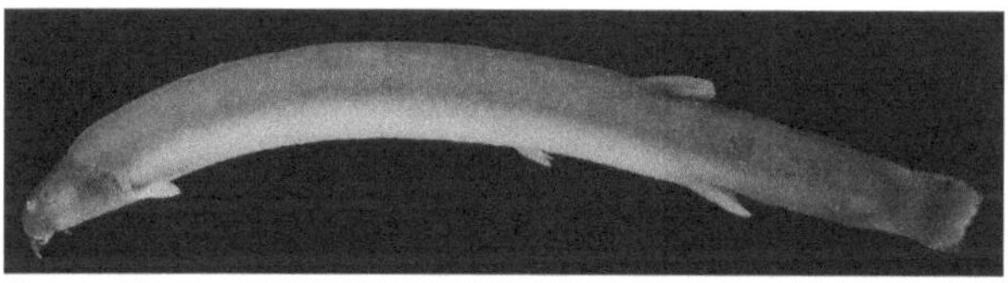

FIG. 100 *Pangio pangia*, RGUMF-397, 50.5 mm SL: lateral view.

Cobitis pangia, Hamilton 1822: 355, 394, type locality: Northeastern Bengal.

Pangio pangia Kottelat, 2013: 183 (valid).

Material examined. RGUMF-397, 50.5–56.6 mm SL, Siang River at Pasighat, East Siang district, Arunachal Pradesh.

Vernacular names in Arunachal Pradesh. *Remum-poda* (in Nyishi & Galo), *Ribii* (in Adi).

Diagnosis. Body plain color, small, slender, and elongated; dorsal fin inserted conspicuously behind pelvic-fin base with 10 soft rays; lower lip interrupted in the middle, conspicuous adipose dorsal and ventral keels on caudal peduncle.

Coloration. In live, body have plain brown color pattern.

Distribution. Known from India (Assam, Manipur, and West Bengal) and Myanmar. In Arunachal Pradesh: known from the Siang drainage (Brahmaputra basin).

Economic importance. The fish is considered as an ornamental fish.

Conservational status: This fish is a widely spread species occurring in the Ganga-Brahmaputra river basin. IUCN status: Least concern (LC).

8. Family: Balitoridae

44. Genus: *Balitora* Gray, 1830

Balitora Gray 1830: pl. 88 (v. 1). Type species: *Balitora brucei* Gray 1830. Type by monotypy.

Body strongly depressed; head and abdomen flattened; eyes small; mouth small, inferiorly arched, with both jaws covered by horny sheath; scales large; a deep preoral groove extending around the corner of the mouth; lips papillated; gill opening extending on the vertical surface of head; dorsal fin inserted slightly ahead of pelvic fin; anterior most pectoral fin rays with muscular pads on ventral surface.

***Balitora brucei* Gray 1830**

Fig. 101.

FIG. 101 *Balitora brucei*, RGUMF-86, 103.6mm SL: (A) lateral, (B) dorsal, and (C) ventral views.

Balitora brucei Gray, 1830, pl. 88, Fig. 1; type locality: Priang River near Cherrapunji, Assam [Meghalaya], India [original type locality: Mountain stream, India].

Balitora brucei: Kottelat, 1998: 491 (description with neotype designation & valid).

Materials examined. RGUMF-86, 10 exs., 100.0–105.0mm SL, Panye River, Lower Subansiri district, Arunachal Pradesh. RGUMF-unregistered, 05 exs., 95.6–101.4mm SL, Senkhi River, Papum Pare district, Arunachal Pradesh.

Vernacular names in Arunachal Pradesh. *Ngocher* (in Nyishi); *Ngobeii* (in Adi).

Diagnosis. Head and body strongly depressed. Head: lateral length (LHL) 19.0%–22.0% SL, maximum width 17.2%–20.5% SL and depth at nape 6.6%–8.0% SL, interorbital width 41.0%–49.0% LHL, caudal peduncle length 16.7–21.4 SL, lateral line with 62–66 pore scales. Fin counts: iii,8½ dorsal-fin rays, ix,12 pectoral rays, iii,5½ anal rays, and ii,9–10 pelvic rays.

Coloration. In live, dorsal parts of body blackish brown with 3–4 predorsal, 2 subdorsal, and 5 post-dorsal dark blotches enclosed by a lighter yellowish margin. Fins darker proximally, yellowish brown with darker markings distally.

Distribution. Known from India: Assam, Meghalaya, West Bengal, and Arunachal Pradesh; Bhutan; Bangladesh; and Nepal. Arunachal Pradesh: see materials examined.

Remarks. *Bhavania arunachalensis* (Nath et al., 2007) was described from the Naodhing River in Arunachal Pradesh. *Balitora brucei* differs from *Bhavania arunachalensis* (data from original description) by its fewer 62–66 (vs. 70–75) lateral line scales and 2 simple (vs. 3 simple) pelvic-fin rays. However, verification of the type specimens of *B. arunachalensis* is required for confirmation of the generic status.

Economic importance. This species has high ornamental value.

Conservational status: This fish is a widely spread in the Ganga-Brahmaputra river Basin. IUCN status: Near threatened (NT).

9. Family: Nemacheilidae

45. Genus: *Aborichthys* Chaudhuri, 1913

Aborichthys Chaudhuri 1913: 244. Type species: *Aborichthys kempi* Chaudhuri 1913. Type by monotypy.

Body elongated, compressed; abdomen rounded; head large and depressed; snout spatulated; mouth terminal; wide; eyes large; a distinct interneural shield in between the occipital process and the basal bone of dorsal fin; barbels four pairs, one each of maxillary, nasal, and two mandibulars; gill membranes free from each other and also from isthmus; rayed dorsal fin inserted ahead of the midway between snout lip and caudal fin base; adipose fin base; adipose dorsal fin low; long; caudal fin forked; lateral line complete.

Key to the species of *Arborichthys*:

1a. Lateral bands indistinct or absent on caudal peduncle .... *Arborichthys* sp

1b. Lateral bands distinctly present throughout the body (including caudal peduncle).......2

2a. Caudal fin margin truncate, 4–5 irregular bands on caudal fin .... *A. waikhomi*

2b. Caudal fin margin rounded, 1–3 bands on caudal fin 3

3a. Lateral line complete..
...............*A. elongatuss*
3b. Lateral line incomplete...4
4a. 15–20 narrow blackish bands along sides ....*A. tikaderi*
4b. 21–24 narrow blackish bands along sides...................................
..............*A. boutanensis*

***Aborichthys boutanensis* (McClelland & Griffith 1842)**
Fig. 102.

FIG. 102 *Aborichthys boutanensis*, RGUMF-435, 58.5 mm SL: lateral view.

Cobitis boutanensis McClelland & Griffith, 1842: 586; type locality: Boutan [Bhutan], on Mishmee Mountains.

Aborichthys cataracta, Arunachalam et al., 2014: 34,

Aborichthys verticauda, Arunachalam et al., 2014: 37

Aborichthys boutanensis: Thoni & Hart, 2015: 292 (valid).

Material examined. RGUMF-435, 02 exs., 58.5–82.9 mm SL, Dikrong River at Doimukh, Papum Pare district, Arunachal Pradesh.

Vernacular names in Arunachal Pradesh. *Remum/Ribreli* (in Nyishi); *Reibi/Dosher* (in Adi); *Riibo* (in Galo).

Diagnosis. Body with 21–24 narrow blackish bands along sides up to the base of caudal fin; caudal fin rounded with 2–3 indistinct grayish bands, and the last subdistal band more prominent than the anterior bands and rounded, a conspicuous rounded spot on the upper side of caudal-fin base present.

Coloration. Body ground color yellowish superimposed with 20–24 blackish bands.

Distribution. Currently known from the Brahmaputra basin in India: Arunachal Pradesh (Dikrong River and Kameng River); and Bhutan.

Remarks. *A. boutanensis* differs from *A. kempi* (a species presently treated as synonym of *A. boutanensis*) by its prominent bands on caudal peduncle (vs. lacking bands on caudal peduncle).

Economic importance. This species has high ornamental value.

Conservational status: This fish is a widely spread species occurring in the Ganga-Brahmaputra river basin. IUCN status: Not assessed (NA).

***Aborichthys elongatus* Hora 1921**
Fig. 103.

FIG. 103 *Aborichthys elongatus*, RGUMF-95, 85.3 mm SL: lateral view.

Aborichthys elongate, Hora 1921: 735, pl. 1, Figs. 1–1b; type locality: Reang River, Darjiling [Darjeeling] District, India, Himalayas, elevation 2000 ft.

Aborichthys elongatus: Kottelat, 2012: 74 (valid).

Materials examined. RGUMF-95, 13 exs., 65–100 mm SL, Senkhi River (a tributary of Dikrong River) in Itanagar, Papum Pare district, Arunachal Pradesh. RGUMF-unregistered, 10 exs., 75.3–100.2 mm SL, Epipani River (a tributary of the Dibang River) at Roing, Lower Dibang valley district, AP.

Vernacular names in Arunachal Pradesh. *Remum/Ribreli* (in Nyishi); *Reibi/Dosher* (in Adi); *Riibo* (in Galo).

Diagnosis. Lateral line complete; vent situated distinctly close to the base of caudal fin than to snout-tip; a prominent blackish ocellus spot present at the upper corner of caudal-fin base; dorsal fin streaked with 2–3 rows of blackish spots; caudal fin dusky with 2–3 grayish bands, distal margin whitish.

Coloration. Body yellowish with black bands all over the body and one at the tip of caudal fin.

Distribution. Known only from India (Darjeeling). In Arunachal Pradesh: known from Dikrong and Dibang River drainage (both Brahmaputra basin).

Economic importance. This species has high ornamental value.

Conservational status: This fish is a widely spread species occurring in the state Arunachal Pradesh. IUCN status: Least concern (LC).

***Aborichthys tikaderi* Barman 1985**

Fig. 104.

FIG. 104 *Aborichthys tikaderi*, RGUMF-409, 92.2 mm SL: lateral view.

Aborichthys tikaderi, Barman, 1985: 680, Fig. 1; type locality: Namdapha Wildlife Sanctuary, Arunachal Pradesh, India.

Aborichthys tikaderi: Kottelat 2012: 74 (valid).

Materials examined. RGUMF-409, 15 exs., 64.6–100.2 mm SL, Lohit River at Tezu, Lohit district, Arunachal Pradesh.

Vernacular names in Arunachal Pradesh. *Remum/Ribreli* (in Nyishi); *Reibi/Dosher* (in Adi); *Riibo* (in Galo).

Diagnosis. Body and caudal peduncle are uniform in depth throughout (body dorsal profile parallel to ventral profile); anus situated close to tip of snout than to base of caudal-fin base (preanal length 48.0–48.2 mm SL); lateral line incomplete (terminating anterior to pelvic-fin origin); body marked with 15–20 black vertical band, and band width equal to the interspace width.

Coloration. Body base color pale olivaceous, superimposed with 15–20 blackish vertical bands which encircle the body leaving a short space on ventral surface. Dorsal surface of head marked with thin blackish curve lines, which inosculate with one another. Body bar width is equal to that of interspaces. Pectoral, pelvic, and anal fins dull white colored. Dorsal fin provided with 2–3 transverse rows of black spots across their rays. Caudal fin light grayish in color, with a semicircular blackish distal marginal band and a black spot near the dorsal margin of caudal-fin base.

Distribution. Known only from the headwater of the Brahmaputra river system in Arunachal Pradesh: Namdapha Wildlife Sanctuary, Kalpangi River at Yachuli (in Lower Subansiri district), Labia River at Dev (in Papum Pare district), Sipu River at Doji (in West Siang District), and Noadhing River (Bagra et al., 2009).

Remarks. They occur in rivers with boulder and pebbly substrates.

Economic importance. This species has high ornamental value.

Conservational status: This fish is a widely spread species occurring in the Brahmaputra river basin only. IUCN status: Vulnerable (VU).

***Aborichthys waikhomi* Kosygin 2012**

Fig. 105.

FIG. 105 *Aborichthys waikhomi*, RGUMF-410, 69.2 mm SL: lateral view.

Aborichthys waikhomi, Kosygin 2012: 49, Figs. 3–4; type locality: Arunachal Pradesh, Bulbulia stream near Bulbulia a tributary of the Noa-Dihing River, Namdapha, India, 27°31′56.4″N, 96°27′32.2″E.

Aborichthys waikhomi: Khynriam & Sen 2014: 8 (valid).

Material examined. RGUMF-410, 10 exs., 65.2–77.5 mm SL, Sinkin River at Anpum, Lower Dibang Valley district.

Vernacular names in Arunachal Pradesh. *Remum/Ribreli* (in Nyishi); *Reibi/Dosher* (in Adi); *Riibo* (in Galo).

Diagnosis. Lateral line incomplete, terminating before pelvic-fin origin; body marked with 12–16 oblique blackish bands narrower than the width of interspaces at the caudal peduncle; a distinct blackish blotch appears near the upper margin of the caudal-fin base; caudal fin truncate, with irregularly shaped bands; dorsal fin with 7 branched rays; and caudal fin with 15–16 branched rays.

Coloration. Body background pale yellow, superimposed with 30–35 bands on flank, head and dorsum marked with several blackish lines which are inosculate with one another. Caudal fin with 4–5 rounded bands. A blackish spot on the dorsal margin of caudal-fin base present.

Distribution. This species is known only from the Brahmaputra drainage in Arunachal Pradesh: Noa-Dihing River in the Namdapha National Park and Tiger Reserve (Changlang district). The authors recorded the species from the Sinkin River (a tributary of the Siang River) in Lower Dibang valley district, Arunachal Pradesh.

Remarks. Known largest size of this species is ca. 68 mm SL.

Economic importance. This species has high ornamental value.

Conservational status: This fish is a widely spread species occurring in the state Arunachal Pradesh. IUCN status: Not assessed (NA).

***Aborichthys* sp.**

Fig. 106.

FIG. 106 *Aborichthys* sp. RGUMF-92–94, 65 mm SL, lateral view.

Material examined. RGUMF-92–94, 38 exs., 38–85 mm SL, Kalpangi River in Lower Subansiri district, Labia River in Papum Pare district and Sipu River at Doji in West Siang, Arunachal Pradesh.

Vernacular names in Arunachal Pradesh. *Remum/Ribreli* (in Nyishi); *Reibi/Dosher* (in Adi); *Riibo* (in Galo).

Diagnosis. Vent distinctly close to caudal-fin base than tip of snout; body with 16–21 black bands, bands indistinct on caudal peduncle.

Coloration. In live, adult body coloration shows a yellowish background superimposed with prominent 16–21 blackish oblique (inclined toward head) bands on sides, except the caudal peduncle, the bands are broader dorsally and tapering downwards and disappears near ventral surface of body. Dorsal fin with 3–4 rows of spots. A distinct blackish spot near the dorsal margin of the base of caudal fin present. Caudal fin with a rounded blackish distal band.

Distribution. Known from the head water of Brahmaputra basin in Arunachal Pradesh: Kalpangi, Siyom, Sipu, and Siang rivers and Abor hills.

Remarks. This fish is a distinct species referable to *Aborichthys kempi* Chaudhuri 1913, described from a stream flowing into the Dihang River (or the Siang River) between Renging and Rotung villages in West Siang district of Arunachal Pradesh. Currently, *A. kempi* is treated as a synonym of *Aborichthys boutanensis* (Thoni and Hart, 2015: 292). The species need resurrection.

Economic importance. This species has high ornamental value.

Conservational status: This fish is a widely spread species occurring in the state of Arunachal Pradesh. IUCN status: Not assessed (NA).

46. Genus: *Acanthocobitis* Peters, 1861

Acanthocobitis Peters 1861: 712. Type species: *Acanthocobitis longipinnis* Peters 1861. Type by monotypy.

Lower lip interrupted in the middle and each side bears a large papillated pad; upper lip with 2–5 rows of papillae and continuous with the lower lip; a conspicuous ocellus marking on upper half of caudal-fin base; caudal fin pointed posteriorly, 17½–19½ branched dorsal-fin rays; more triangular-shaped head in lateral view; vertically oriented suborbital flap; papillated pad on either side of wide medial interruption in lower lip, anus close to anal-fin insertion than to pelvic-fin insertion

***Acanthocobitis pavonacea* (McClelland 1839)**

Fig. 107.

FIG. 107 *Acanthocobitis pavonacea*, RGUMF-387, 93.6 mm SL: (A) lateral view, and (B) mouth part.

Cobitis pavonacea McClelland 1839: 305, 437, pl. 52 (Fig. 1) type locality: Assam, India.

Acanthocobitis pavonacea, Singer & Page 2015: 380 (valid).

Material examined. RGUMF-387, 02 ex., 87.2–93.6 mm SL, Tengapani River at Namsai district, Arunachal Pradesh.

Vernacular names in Arunachal Pradesh. *Remum* (in Nyishi); *Reibi/Dosher* (in Adi); *Riibo* (in Galo).

Diagnosis. Head triangular-shaped in lateral view; upper lip continuous with lower lip; lower lip with a large papillated pad on either side of the medial interruption (Fig. 107B); long dorsal fin with 17½–19½ branched dorsal-fin rays; an ocellus blackish spot with whitish margin located on the upper half of caudal fin base; vent close to anal-fin origin than to pelvic-fin origin, caudal fin with pointed posterior margin.

Coloration. In 10% formalin, 18–20 thin blackish saddles on back descending downwards as bars well below lateral line, small dark spots appears on the upper side between the saddle and bars. An oval-shaped blackish occelous spot with whitish outline present near the dorsal margin of caudal-fin base, 10–12 dark bands on caudal fin.

Distribution. Known from the Ganga-Brahmaputra basin in India and Bangladesh. In Arunachal Pradesh: abundantly encountered in the streams and rivers in Namsai district.

Remarks. This is the only species under the genius of *Acanthocobitis*. Its earlier congener is recently allocated to the genus *Paracanthocobitis*.

Economic importance. This species has high ornamental value.

Conservational status: This fish is a widely spread species occurring in the state Arunachal Pradesh. IUCN status: Not assessed (NA).

47. Genus: *Paracanthocobitis* Grant, 2007

Paracanthocobitis (subgenus of *Acanthocobitis*) Grant 2007: 3. Type species: *Cobitis zonalternans* Blyth 1860. Type by original designation.

Lower lip interrupted in the middle and each side bears a large papillated pad; upper lip with 2–5 rows of papillae and continuous with the lower lip; a conspicuous ocellus marking on upper half of caudal-fin base; emarginate or truncate caudal fin, 9½–15½ branched dorsal-fin rays, rounded head in lateral view; horizontally oriented suborbital flap; papillated pad on either side of narrow medial interruption in lower lip, anus close to to pelvic-fin insertion than to anal-fin insertion.

***Paracanthocobitis botia* (Hamilton 1822)**

Fig. 108.

FIG. 108 *Paracanthocobitis botia*, RGUMF-88, 67 mm SL: lateral view.

Cobitis botia Hamilton, 1822: 350, 394: type locality: Brahmaputra River, Goalpara, India [original type locality: Northeastern Bengal].

Paracanthocobitis botia: Singer and Page, 2015: 386 (Redescription, valid).

Material examined. RGUMF-87–88, 8 exs., 62.0–67.0 mm SL, Poma River at Poma, Sinki River at Itanagar, Papum Pare district, Arunachal Pradesh.

Vernacular names in Arunachal Pradesh. *Remum* (*Remum* (in Nyishi); *Reibi*/*Dosher* (in Adi); *Riibo* (in Galo).in Nyishi); *Reibi*/*Dosher* (in Adi); *Riibo* (in Galo).

Diagnosis. Lateral line complete with 83–100 pores; suborbital flap in male while females have suborbital groove in place of male's suborbital flap; no pelvic lobe or rudimentary; small brownish saddles on back, usually narrower than or equal to the interspace; 8–10 dark blotches along and just below lateral line; 10½–11½ branched dorsal-fin rays; pectoral-fin with 12 rays, and 8 pelvic-fin rays.

Coloration. 8–10 dark oblong to squarish blotches along flank just below lateral line; saddles narrower than or equal to the interspaces, and not extending below to the lateral line. A black roundish spot near dorsal margin of caudal fin base; 5–7 blackish "V"-shaped bands on caudal fin.

Distribution. Currently known only from the Brahmaputra river basin in Assam and Arunachal Pradesh. In Arunachal Pradesh: Kameng, Subansiri, Dibang, Lohit, and Buridhing rivers (Nath and Dey, 2000); and also encountered from Poma, Dikrong, Ranga, Senkhi rivers (see material examined section).

Remarks. Singer and Page (2015) designated the BMNH 1889.2.1.1562, 45.3 mm SL (male), collected from Assam, India as the neotype of *P. botia*.

Economic importance. This species has high ornamental value.

Conservational status: This fish is a widely spread species occurring in the state Arunachal Pradesh. IUCN status: Not assessed (NA).

48. Genus: *Schistura* McClelland, 1838

Schistura McClelland 1838: 944, 947. Type species: Cobitis *rupecula* McClelland 1838. Type by subsequent designation.

Small adult sized with attractive coloration; lower lip medially interrupted without forming two lateral triangular pads; moderately arched mouth, 2.0–3.5 times wider than long; base of caudal fin usually with a black bar or sometimes dissociated; dorsal fin with one or two black marks along its base.

***Schistura walongensis* (Tamang and Sinha 2016)**

Fig. 109.

FIG. 109 *Schistura walongensis*, ZSI/APRC-1190, holotype, 50.5 mm SL: lateral view. (Photograph courtesy of Mr. Lakpa Tamang.)

Physoschistura walongensis Tamang and Sinha 2016: 282, Fig. 1a–d; type locality: small diverted water course of Lohit River (Brahmaputra river basin) at Walong, Arunachal Pradesh, India.

Schistura walongensis: Endruweit, 2017: 383 (valid).

Material examined. ZSI/APRC-1190, holotype, 50.5 mm SL, male; India: Arunachal Pradesh: small diverted water course of Lohit River (Brahmaputra river basin) at Walong; 28°07.395′N, 97°01.096′E.

Vernacular names in Arunachal Pradesh. *Remum/Ribi/Ribreli* (in Nyishi); *Reibo* (in Galo); *Daral* (in Adi).

Diagnosis. Body with 11–18 irregular vertical bars on the sides, more irregular on anterior portion of the body and W-shaped basicaudal bar; broadly rounded suborbital flap in the male; an incomplete lateral line with 71–89 canal pores; axillary pelvic-fin lobe prominent; 37–39 vertebrae; and a forked caudal fin with 9 upper + 8 lower branched rays; dorsal fin with iv,8½ rays, pectoral fin with i,9–10 rays and its second or third branched ray longest.

Coloration. In 10% formalin, body background light yellowish, superimposed with 11–18 irregular vertical brownish bars on flanks, predorsal saddles in the form irregularly scattered overlapping spots, 5–6 post dorsal saddles continuous with lateral bars; ventral surface of head and body light yellowish, dorsal fin with a blackish spot at origin and a blackish brown band along the middle; pelvic, and anal fins hyaline except the proximal portion with light yellowish tinge, caudal fin brownish without any marking.

Distribution. This species is presently known only from its type locality: the Lohit River at Walong, Anjaw district, Arunachal Pradesh.

Remarks. *Schistura walongensis* was originally described under the genus *Physoschistura* and subsequently allocated to *Schistura* by Endruweit (2017). However, the generic placement of this species is uncertain, as its air bladder structure is still not studied.

Economic importance. This species has high ornamental value.

Conservational status: The species is restricted in the stream of Lohit River only in the state Arunachal Pradesh. IUCN status: Not assessed (NA).

***Schistura beavani* (Gunther 1868)**

Fig. 110.

FIG. 110 *Schistura beavani*, RGUMF-91, 62.7 mm SL: (A) lateral view; and (B) mouth part.

Nemacheilus beavani Gunther, 1868: 7: 350; type locality: Kossye River, Uttar Pradesh, India.

Schistura beavani: Kottelat, 2012: 106 (valid). 62.6

Material examined. RGUMF-91, 3 exs. 61.4–66.1 mm SL, Sipu River at Dohi, a tributary of Yomgo River (Siang basin), West Siang district, Arunachal Pradesh (AP).

Vernacular names in Arunachal Pradesh. *Remum/Ribi/Ribreli* (in Nyishi); *Reibo* (in Galo); *Daral* (in Adi).

Diagnosis. Lateral line complete; body ground color yellowish, encircled the body sides and above with 6–7 brownish bands, band width slightly larger than the interspaces width; dorsal fin marked with a transverse row of dark spots in the middle; caudal fin deeply emerginated and its distal half bears two "<" shaped bands, basicaudal bar dissociated (see more characters in remark section).

Coloration. See diagnostic section.

Distribution. Known from India and Nepal. In Arunachal Pradesh: streams and rivers in West Siang district.

Remarks. Some more characters found in the examined specimens are: no suborbital spine in both male and female, processus dentiformis moderately developed, and the corresponding notch on lower jaw very shallow; fin counts: dorsal 8–8½ branched rays; pectoral i,10 rays; pelvic i,6½ rays; anal ii,5½ rays, caudal 9 (upper) + 8 (lower) branched rays.

Economic importance. This species has high ornamental value.

Conservational status: This fish is a widely spread species occurring in the Ganga-Brahmaputra river basin. IUCN status: Least concern (LC)

Schistura zonata **McClelland 1839**

Fig. 111.

FIG. 111 *Schistura zonata*, RGUMF-413, 44.9 mm SL: (A) lateral view (reproduced from McClelland 1839: pl. 53, Fig. 1), (B) lateral, and (C) dorsal views.

Schistura zonata**,** McClelland 1839: 308, 441, pl. 53, Fig. 1; type locality: Upper Assam, India. No types known.

Schistura zonata: Kottelat 2012: 120 (valid)

Material examined. RGUMF-372, 3 exs., 37.4–47.4 mm SL, Dibang River at Roing, Lower Dibang Valley district, Arunachal Pradesh.

Vernacular names in Arunachal Pradesh. *Remum*/*Ribi*/*Ribreli* (in Nyishi); *Reibo* (in Galo); *Daral* (in Adi).

Diagnosis. No suborbital spines in both male and female, body encircled by about 10–12 complete brownish rings, lateral line complete, caudal fin forked, largest adult size ca. 47.4 mm SL.

Coloration. In 10% formalin, body background yellowish, superimposed with 11–12 brownish rings, interspaces broader than body bands.

Distribution. Known from the upper Brahmaputra drainages in Assam and Arunachal Pradesh.

Economic importance. This species is a potential ornamental fish.

Conservational status: This fish is a widely spread species occurring in the Brahmaputra River basin only. IUCN status: Data deficient (DD).

***Schistura devdevi* (Hora 1935)**

Fig. 112.

FIG. 112 *Schistura devdevi*, RGUMF-371, 32.3 mm SL: lateral view.

Nemacheilus devdevi Hora, 1935: 54, pl. 3, Figs. 5 and 6; type locality: Eastern Himalayas, small streams below Darjeeling and Sikkim, India.

Schistura devdevi: Kottelat 2012: 108 (valid).

Materials examined. RGUMF-371, 30 exs., 25.8–34.3 mm SL, Tezu Nallah at Tezu (a tributary of Lohit River), Lohit district, Arunachal Pradesh. RGUMF-unregistered, 55 exs., 25.8–33.9 mm SL, Sinkin River at Anpum, Lower Dibang valley district Arunachal Pradesh. RGUMF-unregistered, 5 exs., 27.2–32.5 mm SL: Kameng River at Tippi, West Kameng district.

Vernacular names in Arunachal Pradesh.

Remum/Ribi/Ribreli (in Nyishi); *Reibo/Riingum* (in Galo); *Daral* (in Adi).

Diagnosis. Lateral line incomplete, terminating above pelvic fin; lacking suborbital spine in both male and female, free posterior gas chamber absent, 5–7 broad bands (= saddles) extending from back to sides of body; basicaudal bar dissociated in the middle; caudal fin lunate. Fin ray counts: dorsal iii,7; pectoral i,8½; pelvic i,5½ (3rd branched longest); anal iii,5½ rays; and caudal i,7 (upper) + 8,i (lower) or i,8 + 7,i.

Coloration. In 10% formalin, body ground color is yellowish, superimposed with 5–7 broad bands (much wider than interspaces). The bands descend from the dorsal surface to the flanks, and extending to well below the midlateral body or may restricted to the upper half of body. Basicaudal bar diffused in the middle frequently. Dorsal and caudal fin are with two transverse rows of grayish spots in the middle.

Distribution. Known from the Ganga-Brahmaputra basin in India and Nepal. In Arunachal Pradesh: Kameng, Lohit, Subansiri, Siang, and Sinkin (a tributary of the Siang River) rivers.

Economic importance. This species has high ornamental value.

Conservational status: This fish is a widely spread species occurring in the Ganga-Brahmaputra river basin. IUCN status: Near threatened (NT).

***Schistura savona* (Hamilton 1822)**

Fig. 113.

FIG. 113 *Schistura savona*, RGUMF-370, 40.1 mm SL: (A) lateral, and (B) dorsal views.

Cobitis savona Hamilton, 1822: 357, 394; type locality: Kosi River at Nathpur, Uttar Pradesh, India (plate 53, Fig. 3, Mc 1839).

Schistura savona: Khynriam & Sen 2014: 25 (valid).

Materials examined. RGUMF-370, 6 exs., 36.6–42.3 mm SL, Sinkin River at Anpum (a tributary of the Siang River, Brahmaputra basin), Lower Dibang Valley district, Arunachal Pradesh. RGUMF-unregistered, 7 exs., 38.3–41.6 mm SL, Kameng River at Tippi, West Kameng district, Arunachal Pradesh.

Vernacular names in Arunachal Pradesh. *Remum/Ribi/Ribreli* (in Nyishi); *Reibo* (in Galo); *Daral* (in Adi).

Diagnosis. Lateral line complete; free posterior chamber of the air bladder absent; body mark with 8–9 narrow yellowish saddles which are descending on flanks, often not reaching lateral line, basicaudal bar triangular in shaped and incomplete; and caudal fin deeply emarginated, adult size reaching up to ca. 42 mm SL. Fins rays count: dorsal iii,9½; pectoral i,10; pelvic i,6 (third branched ray longest); anal iii,5½ rays; and caudal i,9 (upper) + 8,i (lower).

Coloration. In 10% formalin, dorsal half of body dark brown, marked with 9–10 narrow yellowish saddles (4 predorsal, 3 through dorsal fin base, 3–4 post dorsal) extending on sides but not reaching the lateral line; lower half of body yellowish; dorsal fin with a transverse row of black spots in the middle.

Distribution. Known from India: from Kosi River in Uttar Pradesh up to head waters of Brahmaputra River in Arunachal Pradesh; and Nepal. In Arunachal Pradesh: see materials examined section.

Remarks. The *Schistura* sp. reported from Bhutan by Gurung et al. (2013: 4883: Image 4) is most probably *Schistura savona.* Therefore, its distribution range reached Bhutan.

Economic importance. This species is considered as an ornamental fish.

Conservational status: This fish is a widely spread species occurring in the Ganga-Brahmaputra river basin. IUCN status: Least concern (LC).

***Schistura scaturigina* (McClelland 1839)**

Fig. 114.

FIG. 114 *Schistura scaturigina*, RGUMF-386, (A) female lateral view, 44.7 mm SL, and (B) male lateral view, 43.4 mm SL.

Schistura scaturigina McClelland 1839, 1839: 19: 308, 443, Fig. 6, pl. 53; type locality: Assam, India.

Schistura scaturigina: Lalramliana et al., 2014: 211 (valid).

Material examined. RGUMF-386, 8 exs., 34.9–46.3 mm SL, Kameng River at Tippi, West Kameng district, Arunachal Pradesh.

Vernacular names in Arunachal Pradesh. *Remum/Ribi/Ribreli* (in Nyishi); *Reibo* (in Galo); *Daral* (in Adi).

Diagnosis. Lateral line complete; body with 10–12 bars across lateral line, interspaced with narrow short bands from above; caudal fin forked, largest adult size ca. 46.3 mm SL. Fin counts: dorsal iii,8; pectoral i,9; pelvic i,6½; anal ii,5; and caudal i,8 (upper) 7,i (lower).

Coloration. Body with 10–12 bars across lateral line interspaced with narrow short laterally descending bands from saddles above.

Distribution. Known only from India, Nepal, Bangladesh, and Bhutan. In Arunachal Pradesh: known from Kameng River.

Remarks. Suborbital flap in male while females have suborbital groove in the place of male's suborbital flap; axillary pelvic lobe prominent

Economic importance. This species has high ornamental value.

Conservational status: This fish is a widely spread species occurring in the Ganga-Brahmaputra river basin. IUCN status: Least concern (LC).

***Schistura multifasciata* (Day 1878)**

Fig. 115.

FIG. 115 *Schistura multifasciata*, RGUMF-368, 63.3 mm SL: lateral view.

Nemacheilus multifasciatus, Day 1878: 617, pl. 153, Fig. 7; type locality: Darjeeling and Assam, India.

Schistura multifasciata: Lalramliana et al., 2014: 211 (valid).

Material examined. RGUMF-368, 5 exs. 58.5–63.3 mm SL, Siang River at Pasighat, East Siang district, Arunachal Pradesh.

Vernacular names in Arunachal Pradesh. *Remum/Ribi/Ribreli* (in Nyishi); *Reibo* (in Galo); *Daral* (in Adi).

Diagnosis. Lateral line complete; body marked with 14–16 vertical bars which are broader than interspaced, bars anterior to last dorsal-fin ray extremely narrow; caudal fin emarginated or truncated; and dorsal fin with 7½ branched rays.

Coloration. Body background color is yellowish, overlay with 14–16 brownish bars, bars below and in front dorsal fin closely aligned, interspaces narrower than adjacent bars; dorsal and caudal fin spotted by forming 3–4 bands.

Distribution. Known from the Ganga-Brahmaputra basin in India and Nepal. In Arunachal Pradesh: known from the Siang River and its tributaries.

Economic importance. This species is a potential ornamental species.

Conservational status: This fish is a widely spread species occurring in the Ganga-Brahmaputra river basin. IUCN status: Not assessed (NA).

***Schistura tirapensis* Kottelat 1990**

Fig. 116.

FIG. 116 *Schistura tirapensis*, RGUMF-331, 49.4 mm SL: lateral view.

Schistura tirapensis, Kottelat 1990: 118; type locality: Riwa River at Manpong [Nampong], Arunachal Pradesh, India.

Schistura tirapensis: Lokeshwor & Vishwanath 2013: 4 (valid).

Materials examined. RGUMF-331, 2 exs., 45.7–49.4 mm SL, Chatjo River (a tributary of the Burhi-Dihing R., Brahmaputra basin) in Tirap district, Arunachal Pradesh. RGUMF-unregistered, 5 exs., 47.2–56.1 mm SL, Lohit River at Tezu, Lohit district, Arunachal Pradesh. RGUMF-unregistered, 3 exs., 47.2–50.3 mm SL, Hijam River (a small tributary of the Siyom River, Siang basin of Brahmaputra river system), West Siang district, Arunachal Pradesh.

Vernacular names in Arunachal Pradesh. *Remum/Ribi/Ribreli* (in Nyishi); *Reibo* (in Galo); *Daral* (in Adi).

Diagnosis. Lateral line incomplete, ending above the base of pelvic fin; body with 10–12 regular bands separated by yellowish interspaces; dorsal fin with 8 branched rays; caudal forked, adult size reaching up to 56 mm SL.

Coloration. Body marked with 10–12 broad black rings separated by yellowish interspaces, the under surface of body dull white, basicaudal bar dissociated in the middle and a light orange spot appears immediately in front of the dissociated area of the basicaudal bar; a black spot at base of first dorsal rays, distal tip of last simple and subdistal portion of first branched rays orange and a faint grayish transverse mark in the middle of dorsal fin, other fins unspotted, except orange coloration in the middle of dorsal and ventral margin of caudal fin.

Distribution. Currently, known only from Tirap and Lohit districts in Arunachal Pradesh, India.

Remarks. *Noemacheilus arunachalensis* Menon, 1987 has already been preoccupied by *Nemacheilus arunachalensis* Datta & Barman 1984 and become invalid as a junior primary homonym of the later. *Schistura tirapensis* Kottelat (1990) is the replacement name for *Noemacheilus arunachalensis* Menon, 1987.

Economic importance. This species has high ornamental value.

Conservational status: The species is restricted in the Chatjo River a tributary of the Burhi-Dihing R. (Brahmaputra basin) in Tirap district, Arunachal Pradesh. IUCN status: Least concern (LC).

***Schistura rosammae* (Sen 2009)**

Fig. 117.

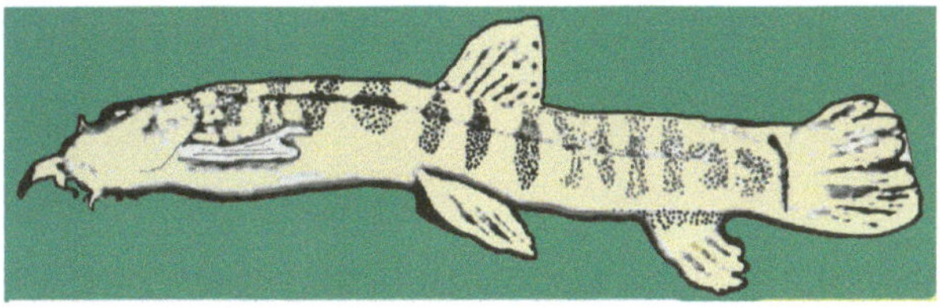

FIG. 117 *Schistura rosammae*, lateral view (redrawn from Sen, 2009: pl. 2A).

Aborichthys rosammai, Sen 2009: 15, pls. 1–2, Fig. 1A and B; type locality: Pabomukh, Subansiri River, Dhemaji district, Assam, India, 27°34′N, 94°14′E, elevation 32 m.

Schistura rosammae: Lalronunga et al., 2013: 589 (valid).

Material examined. No specimen examined (data from Sen, 2009).

Vernacular names in Arunachal Pradesh. *Remum/Ribi/Ribreli* (in Nyishi); *Reibo* (in Galo); *Daral* (in Adi).

Diagnosis. *Schistura rosammae* is distinct in having the following unique combinations of characters: dorsal and ventral profile almost parallel; vent located slightly behind posterior end of pelvic-fin base, and close to the caudal-fin base than to the tip of snout (postanal length: 39.9%–40.6% SL, preanus length 59.4–60.1 mm SL), lateral line complete; dorsal fin with 7½ branched rays; caudal fin more or less truncate.

Coloration. Body base color yellowish, superimpose with 10–11 bands, which coalesced at caudal peduncle region; dorsal surface of head grayish, fins dull white, and a dark band at the base of caudal fin.

Distribution. Known only from the Brahmaputra drainage in Northeast India: Subansiri River in Assam and its upper reaches (Sippi River) in Arunachal Pradesh.

Remarks. Maximum size of this species recorded in the original description is 24 mm SL. In our view, Sen (2009) described the species from the juvenile *Aborichthys* specimens. Members of the genus *Aborichthys* normally grow nearly 100 mm SL to become adulthood. The position of the vent, strongly recommend the species to *Aborichthys*. Ontogenic changes in the body color pattern of all the *Aborichthys* species are unknown, but it is certain that juvenile coloration is quite different from that of the adult.

Economic importance. This species is a potential ornamental fish.

Conservational status: This fish is a widely spread species occurring in the Brahmaputra River Basin only. IUCN status: Not assessed (LC).

***Schistura* sp. 1**

Fig. 118.

FIG. 118 *Schistura* sp. 1: (A) female lateral view, (B) mouth part (RGUMF-89, 72.1 mm SL); (C) male lateral view, and (D) mouth part (RGUMF-90, 75.6 mm SL).

Material examined. RGUMF-89–90, 3 exs., 65.0–90.0 mm SL, Sipu at Doji, and Panye rivers, Papum Pare district, Arunachal Pradesh.

Vernacular names in Arunachal Pradesh. *Remum/Ribi/Reibo ngilang/ Ribreli* (in Nyishi); *Reibo* (in Galo); *Daral* (in Adi).

Diagnosis. No suborbital spine in both male and female, lateral line complete; adult body coloration exhibit 11–14 bars which are wider than interspaces; upper jaw with processus dentiformis and a corresponding deep notch on lower jaw, in male notch on the lower jaw very deeply incised and cheeks inflated; pectoral fin i,10 rays; 8½ branched dorsal-fin rays, 6½ branched pelvic-fin rays, caudal fin with 17 branched rays, scales absent on chest and belly, and caudal-fin emarginate.

Coloration. In 10% formalin, body background pale yellow, encircled the body sides and above with 11–16 vertical dark brown bands, interbands narrower than bars. Fin rays of dorsal fin dark and interadial membrane clear; and other fins hyalin.

Distribution. Known from Sipu & Panye rivers (upper reaches of the Siang River).

Remarks. This fish is likely *Schistura inglisi* (Hora 1935), which is at present regarded as a synonym of *Schistura rupecula* (McClelland 1838) by many authors.

Economic importance. This species is a potential ornamental fish.

Conservational status: The species is restricted in the eastern Himalayas.

***Schistura* sp. 2**

Fig. 119.

FIG. 119 *Schistura* sp. 2, RGUMF-385, 56.1 mm SL: (A) lateral view, and (B) ventral view of mouth part.

Material examined. RGUMF-385, 15 exs., 56.1–75.2 mm SL, Tenga River, a tributary of the Kameng River, West Kameng district, Arunachal Pradesh.

Vernacular names in Arunachal Pradesh. *Remum/Ribi/Ribreli* (in Nyishi), *Reibo* (in Galo), *Daral* (in Adi).

Diagnosis. Lateral line complete, suborbital flap in male, while females have a suborbital groove in place of male's suborbital flap; axillary pelvic lobe prominent; cheeks not inflated in adult males; body with 12–13 irregular bars on flank; caudal fin forked, adult size reaching up to 75.2 mm SL. Fin counts: dorsal ii,7½; Pectoral i,9; anal iii,5½; pelvic i,7; and caudal i,9 + 8,i.

Coloration. In live, overall body is light grayish, superimposed with 10–12 irregular darker (= black) bands, basicaudal bar prominently black, fins without prominent markings.

Distribution. This species occurs in Tenga River, a fast flowing hill stream, in West Kameng district, Arunachal Pradesh.

Remarks. This fish usually caught with *Schizothorax richardsonii*, *Creteuchiloglanish kamengensis*, *Garra* sp., and *Parachiloglanish hodgarti*.

Economic importance. This species has high ornamental value.

Conservational status: The species is restricted in the Tenga River only in the state of Arunachal Pradesh.

49. Genus: *Physoschistura* Bănărescu & Nalbant, 1982

Physoschistura Bănărescu & Nalbant in Singh, Sen, Bănărescu & Nalbant 1982: 208. Type species: *Nemacheilus brunneanus* Annandale 1918. Type by original designation.

A conical free posterior air chamber immediately behind the air-bladder capsule; mouth strongly arched, width 1.5–2.0 times its length; lower lip medially interrupted, forming two broadly triangular lateral pads with a deep furrow; and irregularly shaped bars on the body.

Key to the species of *Physoschistura*:

1a. Lateral line complete, the second branched ray of pectoral fin with a distal filamentous projection, body sides with vertically elongated brownish blotches .... *P. harkishorei*

1b. Lateral line incomplete, no distal filamentous projection in pectoral fin, body sides with vertical bars 2

2a. Body sides with 11–15 irregular dark brown bars which are not tapering ventrally, interspace narrower than bars...*P. dikrongensis*

2b. Body sides with 10–12 vertical bars tapering ventrally, interspace wider than bars.. *Physoschistura* sp.

Physoschistura dikrongensis Lokeshwor & Vishwanath 2012

Fig. 120.

FIG. 120 *Physoschistura dikrongensis*, RGUMF-455, 44.6mm SL: (A) lateral, and (B) dorsal views.

Physoschistura dikrongensis, Lokeshwor & Vishwanath 2012: 250, Figs. 1–4; type locality: Arunachal Pradesh: from Dikarong [Dikrong] River at Doimukh (Brahmaputra basin); 27°08′19″N, 93°44′51″E, India, elevation 120 m.

Physoschistura dikrongensis: Tamang & Sinha, 2016: 287 (valid).

Materials examined. RGUMF-171, 2 exs., 40.0–42.0 mm SL, Dikrong River at Doimukh, Papum Pare district, Arunachal Pradesh (AP). RGUMF-296–301, 6 exs., 43.2–49.1 mm SL, Dikrong River at Doimukh, Papum Pare district, AP, Achom Darshan and party, December 16, 2015. RGUMF-455, 1 ex., 44.6 mm SL, Pare river at Kheel village (a tributary of Dikrong river), Papum Pare district, AP, collected by A. Darshan & A. Santoshkumar, February 10, 2018.

Vernacular names in Arunachal Pradesh. *Remum/Ribi/Ribreli* (in Nyishi); *Reibo* (in Galo); *Daral* (in Adi).

Diagnosis. Lateral line incomplete; body with 11–15 irregular bars, dark brown mottled markings on the head, two V-shaped dark brown bars across the caudal fin toward the distal end; well-formed spherical free posterior chamber of air bladder; dorsal fin with 8½ branched rays; caudal fin forked with 8 + 7 branched rays; suborbital flap in males; and 9 preoperculo-mandibular canal pores.

Coloration. In 10% formalin, body with yellowish cream background superimpose with 11–15 irregular dark brown bars, interspace narrower than bars; 13–14 saddles along back,

Caudal fin hyaline with two V-shaped bars pointing anteriorly. Dorsal fin with three transverse bands and a conspicuous black spot at origin, and hyaline in rest of the fin.

Distribution. Known only from the Dikrong River, Brahmaputra basin in Arunachal Pradesh, India.

Remarks. This is the only species hitherto known under the genus *Physoschistura* from Arunachal Pradesh. The actual diversity of this group of fishes may certainly high in the mountains of the state.

Economic importance. This species has high ornamental value.

Conservational status: The species is restricted in the Dikrong River only (Brahmaputra basin) in the state Arunachal Pradesh. IUCN status: Not assessed (LC).

Physoschistura harkishorei **Das & Darshan 2017**

Fig. 121.

FIG. 121 *Physoschistura harkishorei*, RGUMF-290, holotype, 41.3 mm SL: (A) lateral, and (B) dorsal views.

Physoschistura harkishorei, Das & Darshan 2017: 404, Fig. 1; type locality: India, Arunachal Pradesh, Lower Dibang Valley district, Dibang River (the Brahmaputra basin); 28°09′59″N 95°43′55″E.

Materials examined. RGUMF 290, holotype, 41.3 mm SL; male, India: Arunachal Pradesh state, Lower Dibang Valley district, Dibang River (the Brahmaputra basin); 28°09′59″N 95°43′55″E; Boni Amin Laskar, August 2004. RGUMF 291–295, paratypes, 5 exs., 42.5–53.4 mm SL; same data as above. RGUMF 332, paratype, 1 ex., 46.1 mm SL, male, India: Arunachal Pradesh state, Lohit district, Lohit River at Alubari Ghat, at the immediate side of Lohit bridge (Alubari bridge), 27°51′29″N 96°01′36″E, elevation 159 msl; A. Darshan and Santoshkumar Abujam, August 29, 2016.

Vernacular names in Arunachal Pradesh. *Remum/Ribi/Ribreli* (in Nyishi); *Reibo* (in Galo); *Daral* (in Adi).

Diagnosis. A distinct species of *Physoschistura having the* following combination of characters: the second branched ray of pectoral fin with a distal filamentous projection, body color pattern consisting of 9–10 vertically elongated brownish spots or blotches along flanks, 8–10 brownish saddles on back, saddles not continuous with the lateral blotches; lateral line complete, prominently developed axillary pelvic-fin lobe, well developed free posterior chamber of the air-bladder and caudal fin with 7 + 8 branched rays.

Distribution. This species is abundantly occurring in the Dibang and Lohit rivers of Arunachal Pradesh.

Remarks. *Physoschistura harkishorei* differs from all known congeners by its filamentously projected second branched ray of pectoral.

Economic importance. This species has high ornamental value.

Conservational status: This species is endemic to the upper Brahmaputra basin. IUCN status: Not assessed (LC)

***Physoschistura* sp.**

Fig. 122.

FIG. 122 *Physoschistura* sp., RGUMF-384, 42.9 mm SL: lateral view.

Material examined. RGUMF-384, 03 exs., 38.7–42.9 mm SL, Tezu River at Tezu, Lohit district, Arunachal Pradesh.

Vernacular names in Arunachal Pradesh. *Remum/Ribi/Ribreli* (in Nyishi); *Reibo* (in Galo); *Daral* (in Adi).

Diagnosis. A distinct species of *Physoschistura* having the following combination of characters: lateral line incomplete, extending up to the level through middle of dorsal fin base; suborbital flap in males and suborbital groove in females; body color pattern consisting of 10–12 vertical bars along flanks, 10–13 brownish saddles which are frequently contiguous with lateral bars; prominently developed pelvic-fin lobe.

Distribution. Known from the Tezu river, a tributary of the Lohit River in Lohit district, Arunachal Pradesh.

Economic importance. This fish has high ornamental value.

Conservational status: The species is restricted in the Tezu river, a tributary of the Lohit River (Brahmaputra basin) in the state Arunachal Pradesh.

V. ORDER: SILURIFORMES

The order is distinct by its naked body, barbels usually four pairs (one each of nasal, maxillary and two pairs mandibular) (barbels are used in detecting food), the nasal and chin barbels may be variously absent; adipose fin usually present, body often covers with bony plates; absence of parietal (probably parietals untracebly fused to supraoccipital), symplectic, suboperculum, basihyal and intermuscular bones; mesopterygoid very reduced; relatively small preopercle and interopercle; posttemporal untraceable probably due to fusion with supracleithrum (but thought by some to be present as a separate element in many families); usually with toothed-vomer; medial-radial absent in dorsal and anal fins pterygiophores; dorsal and pectoral fins bears spinous rays in front (the dorsal fin of most catfish usually has two spines: the first shorter one refers to spinelet and the second as the spine); maxilla toothless and small (with exceptions in Diplomystidae and Hypsidoridae); the second third and fourth vertebrae were fused to a single ossification called a "complex vertebrae"; principal rays of the caudal fin fewer than 10+9, with upper principal rays equal to, or fewer than, the lower rays; hypural plates in caudal fin skeleton varying from having six separate hypural plates to complete fusion of all elements; eyes usually small, air-breathing organs developed in Clariidae and Heteropneustidae.

10. Family: Siluridae

50. Genus: *Ompok* Lacepède, 1803

Ompok Lacepède 1803: 49. Type species: *Ompok siluroides* Lacepède 1803. Type by monotypy.

Body compressed, elongated; abdomen rounded; head small; snout rounded, depressed; mouth superior; wide, thin lips, cleft of mouth oblique, not extending to the anterior margin of orbital rim; jaws subequal, lower jaw prominent; teeth villiform on jaws and palate (2 patches in palate); barbels 2 pairs, 1 each of maxillary and mandibular (occasionally absent or rudimentary); rayed dorsal inserted above the middle of pectoral fin; adipose dorsal absent; anal fin long with more than 50 rays; caudal fin forked; lateral line complete.

Key to the species of *Ompok*:

1a. Maxillary barbels shorter than the lateral length head .... *O. pabo*

1b. Maxillary barbels longer than the lateral head length 2
2a. Distal tip of pelvic fin extending to the origin of anal fin......................... .... *O. pabda*
2b. Distal tip of pelvic fin not extending to the origin of anal fin.................. ....*O. bimaculatus*

***Ompok bimaculatus* (Bloch 1794)**
Fig. 123.

FIG. 123 *Ompok bimaculatus*, RGUMF-unregistered, 125 mm SL: lateral view.

Silurus bimaculatus Bloch 1794: 24, pl. 364; type locality: Tranquebar, Tharangambadi, India.

Ompok bimaculatus: Kottelat, 2013: 235 (valid).

Materials examined. RGUMF-108, 2 exs., 75.0–80.0 mm SL, Ranga River, Papum Pare district, Arunachal Pradesh (AP). RGUMF-unregistered, 4 exs., 80.2–130.5 mm SL, Dikrong River at Doimukh, Papum Pare district, AP.

Vernacular name in Arunachal Pradesh. *Pabo* (in Galo), *Beyek/Aalong* (in Adi).

Diagnosis. Jaws subequal and lower jaw prominent; maxillary barbels longer than head; pectoral fin extending beyond origin of dorsal and with a strong serrated spine on its inner edge. Pelvic fins not reaching anal-fin origin and bearing 8 rays. Dorsal fin with i,3 rays; pectoral fin with i,12–14; anal fin with 55–64 rays and caudal fin with 17–18 rays. Lateral line complete.

Coloration. In live, body silvery with purple dorsally faintly greenish dark with a tinged of golden yellow and whitish on the abdomen, a dark blotch behind operculum and above the middle of pectoral fin; occasionally the caudal fin edged with gray and its tips black.

Distribution. India: Ganga-Brahmaputra river system; Bangladesh; Pakistan, Sri Lanka, Myanmar, and Thailand. In Arunachal Pradesh: see materials examined.

Remarks. Currently known *O. bimaculatus* is likely a species complex, consisting of various undescribed species of *Ompok.* We strongly recommend a through taxonomic review on the population in the Chindwin-Irrawaddy, Ganga-Brahmaputra, and Sindh basins.

Economic importance. This species is generally considered as a food fish.

Conservational status: This fish is a widely spread species occurring in the multiple Basins of Southeast Asia. IUCN status: Near threatened (NT).

***Ompok pabo* (Hamilton 1822)**

Fig. 124.

FIG. 124 *Ompok pabo*, RGUMF-109, 170mm SL: lateral view.

Silurius pabo Hamilton, 1822: 153, 375, pl. 22, Fig. 48; type locality: Brahmaputra River, Assam.

Ompok pabo: Kottelat, 2013: 236 (valid).

Materials examined. RGUMF-109, 1 ex., 170mm SL, Dikrong River at Doimukh, Papum Pare district, Arunachal Pradesh (AP); RGUMF-unregistered, 4 exs., 128.7–157.2mm SL, Subansiri River in Lower Subansiri district in AP; RGUMF-unregistered, 2 exs., 132–149.4mm SL, Ranga River in Papum Pare district, AP.

Vernacular name in Arunachal Pradesh. *Pabo* (in Galo), *Beyek/Aalong* (in Adi).

Diagnosis. Body compressed; maximum body depth located at the pelvic-fin origin, dorsal profile gently convex. Mandibular barbels small and not reaching to orbit; maxillary barbels shorter than the head length, and lower border of the orbital rim below level of mouth cleft. Dorsal fin with i,4 rays; pectoral fin with i,14–15 rays; pelvic fin with i,9–10 rays; caudal fin with 18–19 rays. Caudal fin lobes pointed and lateral line complete.

Coloration. In live, body silvery tinged with small spots on dorsally. Light yellowish to gold on upper part of head; whitish to dull silver on belly.

Distribution. Known from the Ganga-Brahmaputra river system in India, Bangladesh, Pakistan, and Myanmar. In Arunachal Pradesh: see materials examined section.

Remarks. Live or fresh specimen samples of this species from the Chindwin-Irrawaddy basin are darker (grayish) in color.

Economic importance. This species is generally considered as food fish.

Conservational status: This fish is a widely spread species occurring in the multiple basins of Southeast Asia. IUCN status: Near threatened (NT).

***Ompok pabda* (Hamilton 1822)**
Fig. 125.

FIG. 125 *Ompok pabda*, RGUMF-400, 134 mm SL: lateral view.

Silurius pabda Hamilton, 1822: 150, 374, pl. 25, Fig. 47; type locality ponds and rivers of Bengal.

Ompok pabda: Gurung *et al.*, 2013: 4883 (valid).

Materials examined. RGUMF-400, 4 exs., 112–134.8 mm SL, Alubari Ghat at Lohit River in Lohit district, Arunachal Pradesh (AP). RGUMF-unregistered, Dikrong at Doimukh, Papum Pare district, AP; RGUMF-unregistered, 6 exs., 87.8–127.5 mm SL, Siang River in Lower Dibang Valley district in AP.

Vernacular name in Arunachal Pradesh. *Pabo* (in Galo), *Beyek/Aalong* (in Adi).

Diagnosis. Snout obtuse, mandibular barbels fairly large and reaching up to the hind edge of the orbit; maxillary barbels longer than head length; pelvic fin originate ahead of the dorsal fin and reaching to the anal-fin origin and anal fin insertion opposite to or one eye diameter distance behind last dorsal-fin ray. Dorsal fin with i,3 rays; pectoral fin with i,11–12 rays; pelvic fin with i,6–7 rays; and caudal fin with 17–19 rays. Lateral line complete.

Coloration. In live, body silvery green, tinged with yellowish to gold and dull silver on belly. A dark spot shoulder on lateral line above the middle of pectoral fin and another close to caudal-fin base.

Distribution. Known from the Ganga-Brahmaputra river system in India; Bangladesh; Pakistan; and Myanmar. In Arunachal Pradesh: see materials examined section.

Economic importance. This species is generally considered as food fish.

Conservational status: This fish is a widely spread species occurring in the multiple basins of Indian ocean. IUCN status: Near threatened (NT).

51. Genus: *Pterocryptis* Peters, 1861

Pterocryptis Peters 1861: 712. Type species: Fem. *Pterocryptis gangelica* Peters 1861. Type by monotypy.

Body elongated and laterally compressed; head flattened and broader than body; eyes small; mouth subterminal, gape horizontal extending to or just beyond vertical to anterior margin of eyes; lower lip thin and papillated; villiform teeth on jaws and palate, vomerine teeth in one or two well-separated small patches; mandibular muscle prominent, cheek inflated. Small or rudimentary dorsal fin. Anal fin fused with caudal fin with a distinct notch in between.

Key to the species of Pterocryptis:

1a. Dorsal fin rudimentary with 2 rays; 13–14 branched pectoral-fin rays, 67–75 branched anal-fin rays..*P. gangelica*
1b. Dorsal fin rudimentary with 1 ray; 11 branched pectoral-fin rays; 85 branched anal-fin rays...*P. indicus*

***Pterocryptis gangelica* Peters 1861**
Fig. 126.

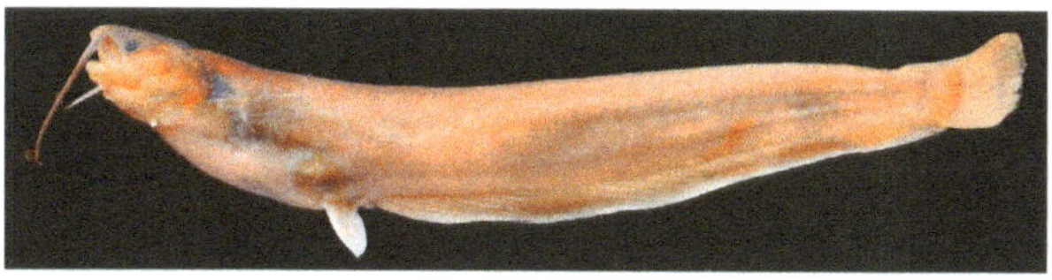

FIG. 126 *Pterocryptis gangelica*, RGUMF-110, 125 mm SL: lateral view.

Pterocryptis gangelica Peters 1861: 712, type locality: Ganges River, India.
Pterocryptis gangelica: Ferraris, 2007: 376 (valid).

Materials examined. RGUMF-110, 6 exs., 100–125 mm SL, Dikrong River at Doimukh, Papum Pare district, Arunachal Pradesh (AP). RGUMF-unregistered, 4 exs., 112.2-122.5 mm SL, Ranga River at Kimin, Papum Pare district, AP.

Vernacular name in Arunachal Pradesh. *Tame* (in Nyishi & Galo); *Bayek* (in Adi).

Diagnosis. Adipose dorsal fin absent. Rayed dorsal fin very short, rudimentary and inserted above half of pectoral fin. Pectoral fin reaching pelvic fin and pectoral spine weakly serrated along the inner edge. Pelvic fin with a very little interspace between it and anal fin origin. Branchiostegal 9; dorsal fin rudimentary with 2 rays; pectoral fin with I,13–14; anal fin long with 67–75 rays; caudal fin without any interspace and rounded with 19 rays.

Coloration. In live, body uniformly purplish black or brownish and dorsal surface and sides of head uniformly brownish. Ventral surfaces of head, abdomen yellow with scattered melanophores; lack of dark patches on dorsal surface and sides of head. Anal fin, caudal, pectoral, and pelvic fins hyaline.

Distribution. Known from the Ganga-Brahmaputra river system in Northeast India. In Arunachal Pradesh: see materials section examined.

Remarks. *P. gangelica* differs from *P. indicus* in having more dorsal-fin rays (2 vs. 1) and less anal-fin rays (67–75 vs. 85).

Economic importance. This species consider as an ornamental and food fish.

Conservational status: This fish is a widely spread species occurring in the Ganga-Brahmaputra river basins. IUCN status: Data deficient (DD).

Pterocryptis indicus **(Datta, Barman and Jayaram 1987)**
Fig. 127.

FIG. 127 *Pterocryptis indicus*, RGUMF-401, 250mm: lateral view.

Kryptopterus indicus Datta, Barman and Jayaram, 1987: 8, 29, Fig. 1; type locality: Hornbill Point, Namdapha River, Namdapha Wildlife Sanctuary, Arunachal Pradesh, India.

Kryptopterus indicus: Ferraris, 2007: 367 (valid).

Materials examined. RGUMF-401, 01 ex., 250.0mm SL, Ranga River at Kimin, Papum Pare district, Arunachal Pradesh. RGUMF-unregistered, 6 exs., 134.0–195.0mm SL, stream at Mebo, East Siang district, AP. RGUMF-unregistered, 5 exs., 95.6–160.2mm SL, Tengapani at Namsai, Namsai district, AP.

Vernacular name in Arunachal Pradesh. *Tame* (in Nyishi & Galo); *Bayek* (in Adi).

Diagnosis. Two pairs of mandibular barbells, adipose dorsal fin absent, dorsal fin rudimentary without spine and inserted above half of pectoral fin. Pectoral fin not reaching pelvic and pelvic fin extending ahead of anal fin origin. Dorsal fin with 1 rays; pectoral I,11 rays; pelvic fin with I,7 rays; anal-fin with iii,85 rays, and caudal fin rounded with 17 rays.

Coloration. In live, dorsal surface and sides of head uniformly yellowish brown; pale brownish on the flanks and thickened integument above the anal fin, with distinct randomly distributed irregular dark patches. Anal fin, caudal, pectoral, and pelvic fins hyaline, with occasional small dark brown spots.

Distribution. Known from the Ganga-Brahmaputra river system in Northeast India. In Arunachal Pradesh: see material examined.

Economic importance. This species consider as an ornamental and food fish.

Conservational status: This fish is a widely spread species occurring in the Ganga-Brahmaputra river basins. IUCN status: Data deficient (DD).

52. Genus: *Wallago* Bleeker, 1851

Wallago Bleeker 1851: 265. Type species: *Silurus muelleri* Bleeker 1846. Type by subsequent designation.

Body elongated and compressed; abdomen rounded; head large, depressed; snout spatulate; eyes with free orbital margin; mouth gape wide, reaching to beyond posterior orbital rim; lower jaw longer, teeth villiform in bands in jaws and patches on palate; barbels two pairs, one each of maxillary and mandibular, maxillary barbel extends beyond pelvic fin; dorsal fin small, inserted above pectoral fin without spine; no adipose dorsal; anal fin long with more than 70 rays; caudal fin deeply forked; lateral line complete.

***Wallago attu* (Bloch and Schneider 1801)**
Fig. 128.

FIG. 128 *Wallago attu*, RGUMF-414, 115 mm SL, lateral view.

Silurius attu Bloch & Schneider 1801: 378, pl. 75; type locality: Malabar, India.

Wallago attu: Kottelat, 2013: 239 (valid).

Materials examined. RGUMF-414, 3 exs., 115.4–122.6 mm SL, Subansiri River, Lower Subansiri district, Arunachal Pradesh (AP). RGUMF-unregistered, 5 exs., 150.0–195.0 mm SL, Siang River at Pasighat, East Siang district, AP. RGUMF-unregistered, 3 exs., 200–305 mm SL, Tengapani River at Namsai, Namsai district, AP.

Vernacular name in Arunachal Pradesh. *Borali* (in Nyishi, Adi, & Galo)

Diagnosis. Jaws subequal and lower jaw slightly longer, lateral line complete and eyes in front of the vertical level of corner of mouth. Dorsal fin with i,4; pectoral fin with i,13–15; pelvic fin with i,7–9 and anal fin with 85–95 rays. Caudal fin forked and upper lobe slightly longer.

Coloration. In 10% formalin, body somewhat darker above and lighter below; usually grayish tinged yellowish along the back and fins sometimes covered with fine dots. A faint yellowish bandpresent along the lateral line.

Distribution. Known from India, Bangladesh, Pakistan, Sri Lanka, and Myanmar. In Arunachal Pradesh: see material examined.

Economic importance. This species is generally considered as food fish.

Conservational status: This fish is a widely spread species occurring in the multiple basins of Indian ocean. IUCN status: Near threatened (NT).

11. Family: Chacidae

53. Genus: *Chaca* Gray, 1831

Chaca Gray 1831: 9. Type species: *Chaca hamiltonii* Gray 1831 (= *Platystacus chaca* Hamilton 1822). Type by monotypy.

Body short, depressed, up to anal fin insertion, strongly compressed and sharply tapering behind, head very large, strongly depressed; mouth subterminal and wide; lips thick and fleshy, jaws subequal: lower jaw prominent; small villiform teeth on jaws; palate edentate; gill membranes free up to pectoral base and joined with isthmus; eyes small; three pairs of barbels, all shorter than head; dorsal fin with one spine and four soft rays; pectoral spine serrated internally;

adipose dorsal confluent with caudal fin; gill rakers absent; caudal fin rounded with a long procurrent dorsal and a shorter procurrent ventral part; lateral line complete.

***Chaca chaca* (Hamilton 1822)**

Fig. 129.

FIG. 129 *Chaca chaca*, RGUMF-141, 155 mm SL: dorsal view.

Platystacus chaca Hamilton, 1822: 140, 374, pl. 28, Fig. 43, type locality: Northeastern Bengal, India.

Chaca chaca: Ng & Kottelat, 2012: 37 (valid).

Materials examined. RGUMF-141, 01 ex., 155 mm SL, Dikrong at Doimukh, Papum Pare district, Arunachal Pradesh (AP). RGUMF-unregistered specimens: 5 exs., 145.2–160.2 mm SL, Tengapani River in Namsai, Namsai district; Lohit River at Tezu, Lohit district, AP.

Vernacular name in Arunachal Pradesh. Not available (in Nyishi, Adi, & Galo).

Diagnosis. Body with a flat, broad head, 6 barbels present and all shorter than the head. Dorsal fin very small with 1 spinelet, 1 spines, and 4 soft rays; pectoral fin with 1 spine and 7–8 rays, pelvic fin with 8 rays, anal fin 8–10 branched rays, and caudal fin with 38–40 rays.

Coloration. In 10% formalin, dorsal surfaces of body dark brownish with irregularly scattered dark spots and ventrally pale. The fins have black spots.

Distribution. Known from the Ganges-Brahmaputra drainages in India, Bangladesh, and Nepal. In Arunachal Pradesh: see material examined.

Remarks. It is known as "frog-fish" and is highly valued in the aquarium trade. Chacids are similar in color to the bottom substrate

Economic importance. This species is a high demand and popular ornamental fish.

Conservational status: This fish is a widely spread species occurring in the Ganga-Brahmaputra river basins. IUCN status: Least concern (LC).

12. Family: Ritidae

54. Genus: *Rita* Bleeker, 1853

Rita Bleeker 1853: 122. Type species: *Rita buchanani* Bleeker 1854 (= *Pimelodus rita* Hamilton 1822). Type by monotypy.

Head large and depressed, snout obtuse, jaw subequal, lips thin, mouth subterminal and moderately wide; teeth molariform or in bands or in patches on premaxillary mandible and vomer; barbels 3 pairs: one each of nasal, maxillary, and mandibular, nasal barbels are very short, maxillary pair extends to operculum, mandibular pair extends to preoperculum; gill membrane free from isthmus, also from each other; adipose fin low and short; pectoral fin with 7–10 rays and a spineserrated on both edges; humeral process large and prominent, exposed, and rugose; anal fin ray short with rays ranging between 8 and 13; caudal fin forked; lateral line complete; vertebrae 34–38.

***Rita rita* (Hamilton 1822)**

Fig. 130.

FIG. 130 *Rita rita*, RGUMF-428, 205 mm SL: lateral view.

Pimelodus rita Hamilton, 1822: 165, 376, pl. 24, Fig. 53; type locality: estuaries of West Bengal, India.

Rita rita: Ferraris, 2007: 106 (valid).

Material examined. RGUMF-428, 01 ex., 205.4 mm SL, Tengapani River at Namsai, Namsai district, Arunachal Pradesh.

Vernacular name in Arunachal Pradesh. *Ritha* (in Nyishi, Adi, & Galo).

Diagnosis. Depressed dorsal-fin tip reaching to the origin of adipose fin, dorsal spine smooth and robust with length 25.6%–41.7% SL, maxillary tooth band width 3.1–4.5 times its length, cleitheral process prominent, and rugose. Pectoral spine long and serrated on both edges with 12–23 retrorse serrae along anterior margin and 20–30 antrorse serrae on posterior margin.

Coloration. In live, body grayish above and on sides, whitish beneath.

Distribution. Known from the Ganga-Brahmaputra basin in India, Bangladesh, and Nepal. In Arunachal Pradesh: this species is encountered in the Noadhing and Tengapani rivers in Namsai district.

Remarks. This species was earlier under the family Bagridae. Nelson et al. (2016) placed this species under the family Ritidae.

Economic importance. This species is considered as an esteemed local food fish.

Conservational status: This fish is a widely spread species occurring in the Ganga-Brahmaputra river basins. IUCN status: Least concern (LC).

13. Family: Ailiidae

55. Genus: *Ailia* Gray, 1830

Ailia (subgenus of *Malapterus*) Gray 1830: pl. 85 (v. 1). Type species: *Malapterus* (*Ailia*) *bengalensis* Gray 1830.

Body and head compressed; mouth terminal; snout overhanging, lips thin, jaws subequal, teeth villiform on jaws and in two separate small patches on palate; eyes small; barbels 4 pairs: one each of maxillary, nasal, and two of mandibular; rayed dorsal fin absent; adipose fin small; pectoral fin with a spine and 13–16 soft rays; pelvic fin rudimentary; anal fin long with more than 40 rays; caudal fin forked; lateral line complete.

Ailia coila (Hamilton 1822)

Fig. 131.

FIG. 131 *Ailia coila*, RGUMF-111, 94.6 mm SL: lateral view.

Malapterurus coila Hamilton 1822: 158, 375; type locality: rivers of Bengal, India.

Ailia coila: Ahmed *et al.*, 2013: 286 (valid).

Materials examined. RGUMF-0111; 6 exs., 50–115 mm SL, Ranga River in Papum Pare district, Arunachal Pradesh (AP). RGUMF-unregistered, 16 exs., 60.2–139.8 mm SL, Siang River in East Siang district, AP.

Vernacular name in Arunachal Pradesh. *Batasi* (in Nyishi & Galo); *Aalong* (in Adi).

Diagnosis. Jaws subequal and upper jaw slightly longer. Dorsal fin absent; four pairs of prominent barbels. Adipose fin small and pelvic fin rudimentary. Pectoral fin with 1 spine, i,14–16 rays; pelvic fin with i,5; anal fin with 72–75 rays and caudal fin with 16–19 rays. Caudal fin often edged black and lower lobe slightly longer.

Coloration. In live, body silvery white, some of the fins tipped gray. Caudal fin often edged with a black tinge.

Distribution. Known from Ganga-Brahmaputra river system of India; Nepal, Bangladesh, and Pakistan. In Arunachal Pradesh: see material examined.

Economic importance. This species consider as an ornamental fish.

Conservational status: This fish is a widely spread species occurring in the multiple basins of Indian ocean. IUCN status: Near threatened (NT).

56. Genus: *Clupisoma* Swainson, 1838

Clupisoma (subgenus of *Silurus*) Swainson 1838: 347, 351, 354. Type species: *Silurus garua* Hamilton 1822. Type by original designation (also monotypic).

Body compressed, elongated; whole of abdominal edge or the part between pelvic fin and vent keeled; mouth subterminal, transverse; its cleft not extending to anterior orbital rim; teeth villiform in bands on jaws and palate; barbels 4 pairs: one pair each of maxillary, nasal, and two of mandibular; rayed dorsal fin with a spine and 6–9 soft rays, inserted above pectoral fin; pectoral fin with a spine serrated internally; a small adipose fin may present; caudal fin forked; lateral complete.

Key to the species of *Clupisoma*:

1a. Maxillary barbel long, extending beyond pectoral fin reaching to pelvic-fin origin............... *C. garua*
1b. Maxillary barbel short, extending up to pectoral fin origin .... *C. montanum*

Clupisoma garua (Hamilton 1822)
Fig. 132.

FIG. 132 *Clupisoma garua*, RGUMF-413, 151 mm SL: lateral view.

Silurius garua Hamilton 1822, 156, 375, pl. 21, Fig. 50; type locality: Rivers of Gangetic provinces, India.

Clupisoma garua: Ferraris, 2007: 357 (valid).

Material examined. RGUMF-413, 01 ex., 151 mm SL, Tengapani River at Namsai, Namsai district, Arunachal Pradesh.

Vernacular name in Arunachal Pradesh. *Gharua* (in Galo); *Aalong* (in Adi).

Diagnosis. Lacking an adipose dorsal fin in adult fishes; nasal barbel not reaching the anterior edge of eye; maxillary barbel extending up to pelvic-fin origin and pectoral-fin tip not extending to pelvic-fin origin. Abdominal edge keeled between the pelvic fin and the vent. Rayed dorsal and pectorals each with a serrated spine at its inner edge and lateral line complete. Dorsal fin with I,7 rays; pelvic fin with i,5 rays and substantially fewer anal fin with iii,26–33 branched rays.

Coloration. In live, body greenish above and silvery below. Head dark dorsally to level of eyes, silvery laterally and ventrally; snout margin pale. Ventral part of head and belly pale. Dorsal fin pale dusky anteriorly, posteriorly rays pale.

Distribution. Known from the Ganga-Brahmaputra system in India; Nepal; Bangladesh; and Pakistan. In Arunachal Pradesh: see material examined.

Remarks. *Clupisoma garua* differs from *C. montanum* by its longer maxillary barbel (extending up to pelvic-fin origin vs. extending up to pectoral-fin origin).

Economic importance. This species consider as an ornamental fish.

Conservational status: This fish is a widely spread species occurring in the multiple basins of Indian ocean. IUCN status: Least concern (LC).

***Clupisoma montanum* Hora 1937**

Fig. 133.

FIG. 133 *Clupisoma montanum*, RGUMF-414, 114 mm SL: lateral view.

Clupisoma montana: Hora 1937: 673; type locality: Teesta River below Darjeeling, India.

Clupisoma montanum: Shrestha 2008: 157 (valid).

Material examined. RGUMF-414, 01 ex., 114.7 mm SL, Tengapani River at Namsai, Namsai district, Arunachal Pradesh.

Vernacular name in Arunachal Pradesh. *Gharua* (in Galo), *Aalong* (in Adi).

Diagnosis. Abdominal edge rounded; maxillary barbel short, extending up to pectoral fin origin; anal fin with ii,39 rays.

Coloration. In live, body greenish above and silvery below. Head dark dorsally to level of eyes, silvery laterally and ventrally; snout margin pale. Ventral part of head and belly pale. Dorsal fin pale dusky anteriorly, posteriorly rays pale.

Distribution. Known from the Ganga-Brahmaputra system. In Arunachal Pradesh: this species is recorded here for the first time from the state.

Economic importance. This species consider as an ornamental fish.

Conservational status: This fish is a widely spread species occurring in the Ganga-Brahmaputra river basins. IUCN status: Not assessed (NA).

57. Genus: *Eutropiichthys* Bleeker, 1862

Eutropiichthys Bleeker 1862: 398. Type species: Masc. *Pimelodus vacha* Hamilton 1822. Type by original designation (also monotypic).

Body elongated, compressed; abdomen rounded; head conical; snout pointed; mouth subterminal, wide, ascending, cleft reaching below orbit or slightly beyond; villiform teeth on jaws and palate; barbels 4 pairs, 1 each of nasal and maxillary, 2 of mandibular; rayed dorsal fin inserted above pectoral fin; adipose dorsal short; pectoral fin with an internally serrated spine; anal fin long with more than 35 rays; caudal fin deeply forked; lateral line complete.

***Eutropiichthys vacha* (Hamilton 1822)**
Fig. 134.

FIG. 134 *Eutropiichthys vacha*, RGUMF-113, 145 mm SL: lateral view.

Pimelodus vacha Hamilton 1822: 196, 378, pl. 19, Fig. 64; type locality: all larger freshwater rivers of Gangetic provinces, India.

Eutropiichthys vacha: Kottelat, 2013: 252 (valid).

Materials examined. RGUMF-113, 01 ex., 145 mm SL, Balijan River in Papum Pare district, Arunachal Pradesh. RGUMF-unregistered, 1 ex., Tengapani River at Namsai, Namsai district, Arunachal Pradesh.

Vernacular name in Arunachal Pradesh. Not available (in Nyishi & Galo), *Aalong* (in Adi).

Diagnosis. Dorsal fin inserted above the middle of pectoral fin. Head length 4.4%–5.4% SL; eye diameter 3.0%–4.5% HL and dorsal fin base 30.1%–30.2% HL. Median longitudinal groove on head and not reaching occipital process base. Adipose fin short and small; pectoral fin reaching pelvic fin base; pelvic fin not reaching anal fin. Dorsal fin with I,7 rays; pectoral fin with I,14–16 rays; pelvic fin with i,5 rays and caudal fin forked with 17 rays. Lateral line complete.

Coloration. In live, body grayish to dull black with a streak of navy blue.

Distribution. Known from India, Nepal, Pakistan, Myanmar, and Thailand. In Arunachal Pradesh: see material examined.

Economic importance. This species consider as food fish.

Conservational status: This fish is a widely spread species occurring in the multiple basins of Southeast Asia. IUCN status: Least concern (LC)

14. Family: Horabagridae

58. Genus: *Pachypterus* Swainson, 1838

Pachypterus Swainson, 1838: 346 et seq. Type species: Masc. *Silurus atherinoides* Bloch 1794. Type by subsequent designation.

Body elongated and compressed; abdomen rounded, head small and depressed, eyes large; snout tapering; mouth wide, ventral, jaws unequal, villiform teeth in jaws and palate, vomero-palatine teeth in the form of two patches, teeth on upper jaw exposed; lip thin, abdomen more or less keeled; barbels 4 pairs: 1 pair each of maxillary and nasal, 2 pairs mandibular; dorsal fin with a spine and 6–7 branched rays; adipose dorsal short, posteriorly free; pectoral fin with a spine and 7 branched rays, spine serrated posteriorly; anal fin long with 24–26 rays, caudal fin forked , lateral line complete.

Pachypterus atherinoides **(Bloch 1794)**
Fig. 135.

FIG. 135 *Pachypterus atherinoides*, RGUMF-112, 63.0 mm SL: lateral view.

Silurus atherinoides Bloch, 1794: 48, pl. 371, Fig. 1, type locality: Tranquebar, Tamil Nadu, India.

Pachypterus atherinoides: Eschmeyer, 2017, Updated April 28, 2017 (valid).

Materials examined. RGUMF-112, 09 exs., 50.0–63.0 mm SL, Balijan River in Papum Pare district, Arunachal Pradesh (AP). RGUMF-unregistered, 15 exs., 54.2–74.8 mm SL, Subansiri River in Lower Subansiri district, AP. RGUMF-unregistered, 05 exs., 51.9–72.6 mm SL, Dikrong River In Papum Pare district, AP.

Vernacular names in Arunachal Pradesh. Not available (in Nyishi & Galo), *Aalong* (in Adi).

Diagnosis. Abdomen more or less keeled. Dorsal and pectoral fin with a spine serrated posteriorly. Lateral line complete. Body with 3–4 stripes formed by blackish spots; pectoral fin extending beyond the pelvic fin. Pectoral fin with i,5 rays; pelvic fin with i,4 rays and anal fin with ii,34–39 rays.

Coloration. In live, body silvery tinged with yellow. Minute blackish spots on the back, flank with three stripes formed by blackish dots. Blackish rounded blotches present at the base of caudal fin and behind the operculum. All fins dull in color.

Distribution. Known from India, Nepal, Bangladesh, and Myanmar. In Arunachal Pradesh: see material examined.

Economic importance. This species consider as an ornamental fish.

Conservational status: This fish is a widely spread species occurring in the multiple basins of Indian ocean. IUCN status: Not assessed (NA).

15. Family: Bagridae

59. Genus: *Batasio* Blyth, 1860

Batasio Blyth 1860: 149. Type species: *Batasio buchanani* Blyth 1860 (=*Pimelodus batasio* Hamilton 1822). Type by original designation.

Body short and compressed; abdomen rounded; head small, conical, and compressed, open pores on sides; mouth inferior, lips fleshy; villiform teeth in jaws and palate, teeth band on lower jaw continuous; barbels 4 pairs: 1 each of nasal and maxillary and 2 mandibulars, all barbels not extending beyond head; gill membranes free from isthmus; rayed dorsal with a spine and 7–8 rays, inserted above the middle of pectoral fin, adipose dorsal fin low; pectoral fin with a spine and 5–6 branched rays, spine serrated posteriorly; caudal fin deeply emarginated; lateral line complete.

Key to the species of *Batasio*:

1a. Body coloration consisting of four to six vertical dark brown bars on light brown ground.. 2

1b. Body with no vertical dark brown bars.. 3

2a. Adipose fin short with length 16.9%–22.2% SL, distal tip of adpressed dorsal fin not reaching the origin of adipose fin.. ..... *B. merianiensis*

2b. Adipose fin long with length 24.5%–25.3% SL, addressed dorsal fin overlap on the anterior portion of adipose fin.. *B. fasciolatus*

3a. Presence of a distinct dark brown elliptical spot below dorsal fin, which overlay above a brownish midlateral stripe; adipose-fin base length 24.5%–25.3% SL... *B. batasio*

3b. Spot below dorsal fin absent, adipose-fin base length 10.0%–17.5% SL............................ 4.

4a. Slender caudal peduncle with depth 5.7%–6.6% SL, short adipose base with 10.0%–12.8% SL, sharper snout in lateral view................................... ....*B. spilurus*

4b. Deeper caudal peduncle with 6.7%–8.2% SL, short adipose base with 14.5%–17.5% SL, snout rounded and bulbous in lateral view...................... .... *B. tengana*

***Batasio fasciolatus* Ng 2006**
Fig. 136.

FIG. 136 *Batasio fasciolatus*, RGUMF-102, 67.0mm SL: lateral view.

Batasio fasciolatus Ng, 2006: 107, Fig. 4; type locality: Market at Malbazar, 26°32′30″N, 88°44′17″E, West Bengal, India.

Batasio fasciolatus: Bagra et al., 2009: 338 (listed, valid)

Materials examined. RGUMF-102, 7 exs., 43.0–83.0mm SL; Panye in Lower Subansiri district, Arunachal Pradesh India; RGUMF-unregistered, 5 exs., 65.1–78.5mm SL; Dikrong River, Papum Pare district, Arunachal Pradesh, India.

Vernacular name in Arunachal Pradesh. *Ngora Poga* (in Nyishi), *Ordek* (in Galo), *Narang* (in Adi).

Diagnosis. Adipose-fin base long with 24.5%–25.3% SL, deep caudal peduncle 10.1%–11.8% SL, and adult coloration exhibit 5–6 vertical dark brown bars on a light brown body.

Coloration. In 10% formalin, body and head brownish, with 5–6 vertical dark brown bars: one passing through the eye; second, on dorsal origin, third at interdorsal, fourth below anterior half of adipose-fin, fifth below the posterior end of adipose-fin base, and the last at the base of caudal fin. All fins hyaline except dorsal with a subdistal dark brown ovoid mark on anterior half.

Distribution. Known from Tista River in West Bengal and from Brahmaputra drainages in Assam and Arunachal Pradesh (see material examined).

Economic importance. This species has high ornamental value.

Conservational status: This fish is a widely spread species occurring in the Ganga-Brahmaputra river basins. IUCN status: Least concern (LC).

***Batasio merienensis* Chaudhuri 1913**
Fig. 137.

FIG. 137 *Batasio merienensis*, RGUMF-399, 58.0mm SL: lateral view.

Macrones merianiensis Chaudhuri, 1913: 253, pl. 9, Figs. 1, 1a and b; type locality: Pond at Mariani Junction, Brahmaputra River drainage, northeastern Assam, India.

Batasio merienensis: Tamang and Sinha, 2014: 5739 (valid).

Material examined. ZSI/V/APFS/P-602, 6 exs., 43.4–57.8 mm SL, Sille River, East Siang district, Arunachal Pradesh, India, 27°52′38″N & 95°18′18″E.

RGUMF-399, 2 exs., 50.0–75.9 mm SL, Ramsikri River (a tributary of the Siang River), East Siang district, Arunachal Pradesh, India.

Vernacular names in Arunachal Pradesh. *Tengra* (in Nyishi), *Ordek* (in Galo), *Narang* (in Adi).

Diagnosis. Adipose-fin base length with 16.9%–22.2% SL, adpressed dorsal fin not reaching the anterior origin of adipose fin, body depth at anus 15.2%–18.4% SL, caudal peduncle depth 9.7%–11.5% SL, eye diameter 18.3%–25.9% HL, four vertical dark brown bars on head and body, and no stripe on the body.

Coloration. In 10% formalin, head and body brownish. Four vertical dark brown bars: one passing through the eye; second, on dorsal origin, third at adipose origin and another fourth at base of caudal fin. Tympanum slightly darker than other surface of body. Dorsal fin with brownish subdistal band, pectoral, pelvic and anal fins hyaline.

Distribution. Known from the Brahmaputra River drainage in Assam and Arunachal Pradesh (see material examined).

Remarks. This species is very similar to *B. fasciolatus* in terms of body coloration, but it can be easily differentiated by its shorter adipose-fin base length (16.9%–22.2% SL vs. 24.5–25.3).

Economic importance. This species has high ornamental value.

Conservational status: This fish is a widely spread species occurring in the Brahmaputra River Basin only. IUCN status: Not assessed (NA).

***Batasio batasio* (Hamilton 1822)**

Fig. 138.

FIG. 138 *Batasio batasio*, RGUMF-319, 57.4 mm SL: lateral view.

Pimelodus batasio Hamilton, 1822: 179, 377; type locality: Teesta River, North Bengal, India.

Batasio batasio: Darshan et al., 2011: 57 (valid).

Materials examined. RGUMF-319,1 exs., 57.4 mm SL, Siang River at Pasighat, East Siang district, Arunachal Pradesh (AP). RGUMF-unregistered, 51.1–55.0 mm SL. Sile River (tributary of the Siang River), East Siang district,

Arunachal Pradesh. RGUMF-unregistered, 45.1–57.0 mm SL, 10 exs., Tezu River at Tezu, Lohit district, Arunachal Pradesh. RGUMF-74.4 mm SL. Sinkin River at Anpum (a tributary of the Siang River), Lower Dibang Valley district, Arunachal Pradesh.

Vernacular name in Arunachal Pradesh. *Tengra* (in Nyishi), *Ordek* (in Galo), *Narang* (in Adi).

Diagnosis. Body with a longitudinal grayish midlateral stripe and another shorter stripe above extending from slightly above tympanum to anterior half of the adipose fin. An elliptical dark brown spot on the midlateral stripe below the dorsal fin base. Dorsal spine height with 16.7%–18.5% SL, adipose-fin base length 24.8%–26.5% SL, and body depth at anus 17.0%–17.4% SL.

Coloration. See diagnosis.

Distribution. India (West Bengal, Assam, Tripura, and Arunachal Pradesh), Nepal, and Bangladesh. In Arunachal Pradesh, this species is abundantly occurring in the Dikrong, Ranga, Senkhi, Siang, and Lohit rivers.

Remarks. This species differs from all known congeners (except *Batasio macronotus* Ng & Edds) in the Ganga-Brahmaputra basin by its two parallel stripes on the flank. *B. batasio* differs from *B. macronotus* by its longer dorsal spine (16.7–18.5 vs. 14.0%–16.1% SL), shorter adipose-fin base (24.8–26.5 vs. 28.3%–31.8% SL), and slender body (17.0–17.4 vs. 19.5%–21.2% SL).

Economic importance. This species has high ornamental value.

Conservational status: This fish is a widely spread species occurring in the Ganga-Brahmaputra river basins. IUCN status: Least concern (LC).

***Batasio tengana* (Hamilton 1822)**

Fig. 139.

FIG. 139 *Batasio tengana*, MUMF-unregistered, lateral view. (Photograph courtesy of Prof. Waikhom Vishwanath.)

Batasio tengana Hamilton, 1822: 176, pl. 39, Fig. 58; type locality: Tista River at Tista barrage; 26°45′10″N; 88°34′11″E, West Bengal, India [originally in Hamilton, 1822: Goalpara, Assam, India].

Batasio tengana: Ng, 2006: 102 (redescription, valid).

Material examined. RGUMF-unregistered, 1 ex., 56.1 mm SL, Kameng River, West Kameng district, Arunachal Pradesh.

Vernacular name in Arunachal Pradesh. *Tengra* (in Nyishi & Adi), *Ordek* (in Galo), *Narang* (in Adi).

Diagnosis. Adipose-fin base short with 14.5%–17.5% SL, pectoral fin rays not extending to the pelvic origin, adult size very small usually less than ca. 56.2 mm SL, caudal peduncle depth 6.7%–8.2% SL, and a grayish mid-dorsal stripe present.

Coloration. In live, ground color of head and body yellowish. Occipital region of head brownish. A brownish stripe runs on either side of the dorsal midline of the body, from the occipital region to the adipose fin origin. Dorsal-fin origin with a dark brown patch. A distinct dark brown mid-dorsal stripe, running from the base of last dorsal fin ray to the origin of dorsal procurrent caudal rays. A faint semi-lunate mark appears on the dorsal third of tympanum. All fins generally hyaline.

Distribution. This species is known from the Tista River (upper reaches the Ganges) and the Brahmaputra drainage in Assam and Arunachal Pradesh.

Remarks. Ng (2006) redescribed *B. tengana* and designated a neotype (UMMZ 244796, 43.4 mm SL) to a specimen collected from the Tista River at Tista barrage, West Bengal, India.

Economic importance. This species has high ornamental value.

Conservational status: This fish is a widely spread species occurring in the Ganga-Brahmaputra river basins. IUCN status: Least concern (LC).

***Batasio spilurus* Ng 2006**

Fig. 140.

FIG. 140 *Batasio spirulus*, RGUMF-317, 54.8 mm SL: (A) dorsal, and (B) lateral views.

Batasio spilurus Ng: 2006: 110, Fig. 5; type locality: Assam, Dibrugarh district, about 27°29′N, 94°54′E, India.

Batasio spilurus: Darshan et al., 2011: 52 (valid).

Materials examined. RGUMF-317, 01 ex., 54.8 mm SL; RGUMF-318, 1 exs., 54.1 mm SL; India: Arunachal Pradesh, East Siang district, Siang River at Pasighat, collected by Achom Darshan and Party, February 21, 2015.

Vernacular names in Arunachal Pradesh. *Tengra* (in Nyishi), *Ordek* (in Galo), *Narang* (in Adi).

Diagnosis. Extremely taper snout in lateral view, adipose-fin base very short with length 10.0%–12.8% SL, slender caudal peduncle 5.7%–6.6% SL, body depth at anus 11.9%–14.1% SL, head width 14.7%–16.5% HL, presence of a midlateral blackish line, one blackish spot each on the tympanum and the base of caudal fin.

Coloration. In 10% formalin: head and body yellowish. Skin over occipital region grayish brown color. Dorsal half of tympanum with a semilunate dark brown spot and another triangular blackish mark at the base of caudal fin.

A pale grayish midlateral line extending from the dorsal margin of tympanum to base of caudal fin, and numerous chevron-shaped marks interspersed at regular intervals along midlateral line. Anterior half of dorsal fin with a dark brown ovoid mark.

Distribution. Known from the Brahmaputra River in Dibrugarh district of Assam and also from the Siang River (tributary of the Brahmaputra) in Arunachal Pradesh.

Remarks. Darshan et al. (2017) recorded this species for the first time from the Siang River of Arunachal Pradesh.

Economic importance. This species have high ornamental value.

Conservational status: This fish is a widely spread species occurring in the Brahmaputra river basin only. IUCN status: Not assessed (Na).

60. Genus: *Chandramara* Jayaram, 1972

Chandramara Jayaram 1972: 816. Type species: Fem. *Pimelodus chandramara* Hamilton 1822. Type by original designation (also by monotypy and by absolute tautonymy).

Body short and compressed; head small; mouth subterminal; jaws subequal, teeth on jaws and palate in broad villiform bands;barbels 4 pairs: 1 each of maxillary and nasal, 2 of mandibular, all barbels not extending beyond head; pores on side and ventral surface of head; dorsal fin inserted above the distal tip of cleitheral process with a spine and 7 branched rays; pectoral spine strong serrated internally; anal fin 13–17 rays; caudal-fin forked, lateral line complete; body spotted with numerous black dots.

***Chandramara chandramara* (Hamilton 1822)**
Fig. 141.

FIG. 141 *Chandramara chandramara*, RGUMF-325, 40.9 mm SL: lateral view.

Pimelodus chandramara, Hamilton, 1822: 162, 375; type locality: Atreyi [Atrai] River, North Bengal, India.

Chandramara chandramara, Ferraris 2007: 86 (valid).

Materials examined. RGUMF-325, 01 ex., 40.9 mm SL, Purchased from a local Naharlagun Market, Papum Pare District, Arunachal Pradesh. RGUMF-220, 15 exs., 35.1–47.1 mm SL, Singen River (upper reaches of Simen River, a tributary of the Brahmaputra River), East Siang district, Arunachal Pradesh.

Vernacular names in Arunachal Pradesh. *Tengra* (in Nyishi & Galo), *Narang* (in Adi).

Diagnosis. Body with irregularly scattered blackish spots, orbital margin visible when viewed ventrally, interdorsal equal or shorter than dorsal-fin base length, dorsal fin I,6 rays, pectoral spines long and slender with 11–13 antrorse serrae posteriorly, lateral line complete. Caudal fin forked with equal lobes.

Coloration. Head and body pale olive, with numerous irregularly scattered blackish spots, ventrally pale. A black shoulder spot present. Skin over the snout tip and occiput darker due to the accumulation of melaniphore pigments.

Distribution. Known from India and Bangladesh. Arunachal Pradesh: see material examined section.

Remarks. Some authors treated this species as *Rama chandramara.*

Economic importance. This species has high ornamental value.

Conservational status: This fish is a widely spread species occurring in the Ganga-Brahmaputra river basins. IUCN status: Least concern (LC).

61. Genus: *Mystus* Scopoli, 1777

Mystus Scopoli (ex Gronow) 1777: 451. Type species: Masc. *Bagrus halepensis* Valenciennes 1840. Type by subsequent designation.

Body short to moderately elongated; abdomen rounded; head moderate, upper surface with median longitudinal groove; mouth subterminal; lips thin, jaws subequal, teeth villiform in bands on jaws and palate; barbels 4 pairs: 1 each of nasal and maxillary, 2 of mandibulars, maxillary barbel extended beyond head; gill membranes free from isthmus and also from each other; dorsal inserted above the middle of pectoral fin ray with a spine and 7 rays; pectoral fin spine internally serrated; adipose dorsal fin present; caudal fin forked, lobed unequal; lateral line complete.

Key to the species of *Mystus*:

1a. Long adipose fin, originating immediately behind the base of last dorsal-fin ray and extending to the middle of caudal peduncle..2

1b. Short adipose fin, originated considerably well behind dorsal fin and extendingtothemiddleofcaudalpeduncle...3

2a. No lateral longitudinal band, presence of a dark brown nuchal (in front of dorsal fin) and a humeral (above the pectoral fin) spots ....*M. cavasius*

2b. A light longitudinal band one above and another below lateral line, lower band twice as broad as upper, nuchal or humeral spot absent... *M. bleekeri*

3a. Median longitudinal groove on head reaching base of occipital process, adult size small, that is, mature at about 44 mm SL....................................... ....*M. carcio*

3b. Median longitudinal groove on head not reaching base of occipital process adult size larger, that is, mature at about 75 mm SL.............................. ....4a

4a. Body with a dark gray tympanic spot and another spot at the base of caudal fin, the two spots are interconnected by a thin narrow blackish midlateralline.. *M. dibrugarensis*
4b. Tympanic spot present but the spot at caudal-fin base and the thin narrow blackish midlateral line absent...*M. tengara*

***Mystus cavasius* (Hamilton 1822)**
Fig. 142.

FIG. 142 *Mystus cavasius*, RGUMF-103, 78.2 mm SL: lateral view.

Pimelodus cavasius Hamilton 1822: 203, 379, pl. 11, Fig. 67; type locality: larger freshwater of Gangetic provinces, India.

Mystus cavasius: Ng & Pethiyagoda, 2013: 167 (valid).

Materials examined. RGUMF-103, 6 exs.; 75–100 mm SL, Dikrong River at Doimukh, Papum Pare district, Arunachal Pradesh (AP). RGUMF-unregistered, 78.2–95.1 mm SL, Siang River at Pasighat, East Siang district, AP.

Vernacular names in Arunachal Pradesh. *Ngora-poga* (in Nyishi & Galo), *Tengil* (in Adi).

Diagnosis. Adipose-fin based long commencing immediately from the base of last dorsal-fin ray and extended up to the middle of the caudal peduncle, posterior fontanel reaching to the base of occipital process, body lacking lateral stripes, the presence of a faint blackish spot at the dorsal-fin origin and another humeral spot; dorsal spine short and feebly serrated posteriorly, dorsal fin tall with height almost equal to lateral head length, first branchial arch with 13–22 gill rakers.

Coloration. In live, dorsal surface of head and body, light gray with bluish and golden reflection; flank silvery and lacking lateral stripes; belly whitish; and presence of a small blackish triangular nuchal spot at dorsal-fin origin and another small semi-ocellate humeral spot above the pectoral fin.

Distribution. Known from India, Nepal, Bangladesh, and Pakistan. In Arunachal Pradesh: see material examined.

Remarks. Earlier records of *Mystus cavasius* from Myanmar, the peninsular India and Srilanka are misidentification of *Mystus falcarius*, *Mystus seengtee*, and *Mystus zeylanicus*, respectively.

Economic importance. This species has both ornamental and food value.

Conservational status: This fish is a widely spread species occurring in the Ganga-Brahmaputra and Indus river basins. IUCN status: Least concern (LC).

***Mystus bleekeri* (Day 1877)**
Fig. 143.

FIG. 143 *Mystus bleekeri*, MUMF-9521, 108.3 mm SL: lateral view.

Macrones bleekeri Day 1877: 451, pl. 101, Fig. 1, type locality: Jumna [Yamuna] River, India.

Mystus bleekeri: Kottelat, 2013: 263 (valid).

Materials examined. RGUMF-105; 4 exs., 85.0–110 mm SL, Dikrong River at Doimukh, Papum Pare district, Arunachal Pradesh (AP). RGUMF-unregistered, 64.3–70.7 mm SL, Sijosa, East Kameng district, AP. MUMF-9521, 01 ex., 108.3 mm SL.

Vernacular names in Arunachal Pradesh. *Ngora-poga* (in Nyishi & Galo), *Tengil* (in Adi).

Diagnosis. Posterior fontanel reaching to the base of occipital process; adipose-fin base long, commencing immediately from the base of the last dorsal-fin ray and extending up to the middle of the caudal peduncle; dorsal-fin spine smooth on both edges; body with two light longitudinal bands: one above and one below the lateral line, and gill rakers on the first arch with 11–15.

Coloration. The three broad stripes: dorsal, midlateral, and ventrolateral, are separated by narrow pale interspace. A large tympanic spot present above the pectoral fin.

Distribution. Known from the Ganga-Brahmaputra basin in India, Nepal, and Bangladesh; and Pakistan.

Remarks. Day (1877) described *Mystus bleekeri* based on the type series collected from various river systems viz. the Sind in Pakistan, Yamuna and upper-waters of Ganga in India and Irrawaddy in Burma. Ng and Kottelat (2009) clarified the identity of *M. bleekeri* and restricted its distribution to the Ganga-Brahmaputra basin while the Burmese specimens were referred to *Mystus rufescens*.

Economic importance. This species has both ornamental and food value.

Conservational status: This fish is a widely spread species occurring in the Ganga-Brahmaputra and Indus River basins. IUCN status: Least concern (LC).

***Mystus tengara* (Hamilton 1822)**

Fig. 144.

FIG. 144 *Mystus tengara*, RGUMF-104, 57.0 mm SL: lateral view.

Pimelodus tengara Hamilton 1822: 183, 377, pl. 3, Fig. 61, type locality: Ponds of India.

Mystus tengara: Darshan et al. 2011: 2177 (redescription, valid).

Materials examined. RGUMF-104, 6 exs., 47.0–75.0mm SL, Dikrong River at Doimukh, Papum Pare district, Arunachal Pradesh (AP). RGUMF-unregistered, 15 exs., 55.0–77.5mm SL, Subansiri River, Lower Subansiri district, AP. RGUMF-unregistered, 25 exs., 65.0–72.1mm SL, Tengapani River at Namsai, Namsai district, AP.

Vernacular names in Arunachal Pradesh. *Ngora-poga* (in Nyishi & Galo), *Tengil* (in Adi).

Diagnosis. Posterior fontanel of mature individuals not extending to the base of occipital process, origin of adipose fin located far behind the dorsal fin, adipose-fin base length 24.0%–31.7% SL, 31–42 gill rakers on the first branchial arch, total vertebrae 34–37 (see the coloration section for additional diagnostic features).

Coloration. In 10% formalin, lateral view of the body shows four brown stripes (a mid-dorsal and three lateral stripes), all the stripes are separated by pale longitudinal lines. The pale longitudinal lines separating the mid-dorsal and the lateral stripes originate from below the anterior-base of dorsal fin and extending up to the posterior portion of the adipose-fin base. Lateral lines appear as a thin pale line in the middle of the midlateral stripe. A distinct oval dark brown tympanic spot present above pectoral fin.

Distribution. Known from the Ganga-Brahmaputra River drainage in India, Bangladesh and Nepal. The species is also recorded from Narmada and Mahanadi basins in northern India, Indus River drainage of Pakistan.

Remarks. In juvenile specimens, posterior fontanel may extend up to the base of occipital process. Earlier records of *Mystus vittatus* from Arunachal Pradesh (e.g., Bagra et al., 2009; Gurumayum et al., 2016) are probably due to misidentification of *M. tengara* or *M. carcio*.

Economic importance. This species has both ornamental and food value.

Conservational status: This fish is a widely spread species occurring in the Ganga-Brahmaputra River Basins. IUCN status: Least concern (LC).

***Mystus carcio* (Hamilton 1822)**

Fig. 145.

FIG. 145 *Mystus carcio*, RGUMF-376, 45.2mm SL; lateral view.

Pimelodus carcio Hamilton 1822: 181, 377, pl. 23, Fig. 60; type locality: ponds of northern Bengal, India.

Mystus carcio: Darshan et al., 2010: 49 (redescription, valid).

Materials examined. RGUMF 376, 5 exs., 31.2–45.2 mm SL, Dikrong River at Doimukh, Papum Pare district, Arunachal Pradesh (AP). RGUMF-unregistered, 10 exs., 35.0–47.5 mm SL, Subansiri River, Lower Subansiri district, AP.

Vernacular names in Arunachal Pradesh. *Ngora-poga* (in Nyishi & Galo), *Tengil* (in Adi).

Diagnosis. Small adult size, maturing at about 44.0 mm SL; adipose-fin base length (8.5%–11.9% SL) shorter than or equal to the dorsal-fin base; posterior fontanel reaching to the base of the occipital process, distal tip of adpressed pelvic fin reaching to the origin of anal fin; pectoral girdle with a coracoid shield, exposed ventrolaterally below pectoral fin; eyes large with diameter 25.6%–30.7% HL, vomerine tooth band interrupted at the middle.

Coloration. In 10% formalin, body ground color amber, superimposed with four brownish broad stripes viz. mid-dorsal, dorsolateral, midlateral, and ventrolateral, all separated by pale interspaces. The midlateral stripe originates immediately behind the posterolateral corner of the head, dorsolateral stripes from the region between the epiotic and the base of occipital process, and the ventrolateral stripes from the distal-tip of cleithral process; all the stripes extended continuously up to the base of caudal fin. A distinct dark brown spot is also present on tympanum.

Distribution. Known from the Brahmaputra drainage in the states of West Bengal, Assam, Tripura, and Arunachal Pradesh in India; and also from the Meghna River drainage in Bangladesh. This species was also encountered from the Ganga River at Gay Ghat in Patna, Bihar, India by the first author in 2011 during a field survey.

Remarks. Darshan et al. (2017) recorded this species from Arunachal Pradesh. *M. carcio* can be differentiated from the juveniles of *M. tengara* by its shorter adipose fin, which equal to the dorsal-fin base length (vs. adipose-fin base length longer than dorsal-fin base).

Economic importance. This species has ornamental value.

Conservational status: A widely distributed species in Northeast India and Bangladesh. This species is most abundant in beels in Assam and it is also occurring in foothill streams in Arunachal Pradesh. IUCN status: Least concern (LC).

***Mystus dibrugarensis* (Chaudhuri 1931)**

Fig. 146.

FIG. 146 *Mystus dibrugarensis*, RGUMF-374, 65.5 mm SL: lateral view.

Macrones montanus var. dibrugarensis Chaudhuri, 1931: 254, pl. 9, Fig. 2, 2a and 2b; type locality: Dibrugarh, Assam, India.

Mystus dibrugarensis: Darshan et al., 2011: 2177 (valid).

Materials examined. RGUMF-172, 10 exs., 45.0–80.0 mm SL, Dikrong River at Doimukh, Papum Pare district, AP. RGUMF-374, 12 exs., 54.9–65.5 mm SL, Ramsikri River, a tributary of the Siang River, Lower Dibang Valley district, Arunachal Pradesh (AP). RGUMF-375, 10 exs., 57.9–80.4 mm SL, Sinkin River at Anpum, a tributary of Siang River, Lower Dibang Valley district, AP. RGUMF-unregistered, 55 exs., Dibang River at Roing, Lower Dibang Valley district, AP.

Vernacular names in Arunachal Pradesh. *Ngora-poga* (in Nyishi & Galo), *Tengil* (in Adi).

Diagnosis. Body with two blackish spots: one on the tympanum and the other at the base of caudal fin, the two spots are connected by a thin narrow blackish line originated from the dorsal rim of the tympanic spot; posterior fontanel not extending to the base of the occipital process; adipose fin short and its origin located after a long interdorsal gap; gill rakers 7 + 21 = 28 on the first branchial arch, and total vertebrae 36.

Coloration. In live, dorsal surface of head and body grayish, gradually lighter below, flank light gray with a blackish tyampanic spot and another black circular blotch at the base of caudal fin, the two spots are interconnected by a thin narrow blackish line running along the lateral line, barbels blackish (except the inner mandibular).

Distribution. Known from the headwater of the Brahmaputra River drainage in Assam and Arunachal Pradesh.

Remarks. Reports of *Mystus montanus* from Arunachal Pradesh (e.g., Bagra et al., 2009, Gurumayum et al., 2016) are probably misidentification of *M. dibrugarensis*.

Economic importance. This species has high ornamental value.

Conservational status: This species is endemic only to the upper Brahmaputra basin. IUCN status: Least concern (LC).

62. Genus: *Sperata* Holly, 1939

Sperata Holly 1939: 143. Type species: Fem. *Bagrus lamarrii* Valenciennes 1840. Type by being a replacement name.

Body elongated, compressed; snout broad, spatulated or rounded; mouth subterminal, moderately wide; jaws subequal, lips tin; villiform teeth on jaws and palate; barbels extend posteriorly to pelvic fins or beyond to anal fin; dorsal spine weakly serrated posteriorly; adipose fin base short, about as long as the rayed dorsal-fin base or slightly greater; brownish-gray on back, silvery on flanks and belly. A dark well-defined spot is on the adipose dorsal fin. Caudal fin deeply forked.

Key to the species of *Sperata*:

1a. Snout rounded, adipose-fin base length longer than the dorsal-fin base, 18–20 gill rakers on the first branchial arch.. .... *S. aor*

1b. Snout truncate, adipose-fin base length shorter than the dorsal-fin base, 13–15 gill rakers on the first branchial arch... .... *S. seenghala*

Sperata aor **(Hamilton 1822)**
Fig. 147.

FIG. 147 *Sperata aor*, RGUMF-unregistered; 275 mm SL: lateral view.

Pimelodus aor Hamilton, 1822: 205, 379, pl. 20, Fig. 68; type locality: rivers of Bengal and upper parts of Gangetic estuaries, India.

Sperata aor: Ahmed et al., 2013: 286 (valid).

Materials examined. RGUMF-173; 07 exs., 125–378 mm SL, Subansiri at Gerukamukh, Lower Subansiri district, Arunachal Pradesh (AP). RGUMF-unregistered, 02 exs., 152–275 mm SL, Dikrong River at Doimukh, Papum Pare district, AP.

Vernacular names in Arunachal Pradesh. *Ari* (in Nyishi, Adi, & Galo).

Diagnosis. Head long and dorso-ventrally flattened; adipose-fin base length longer than the dorsal-fin base; snout rounded; an ovoid interneural shield present between the tip of occipital process and the basal bone of dorsal fin; and 18–20 gill rakers on the first branchial arch.

Coloration. In live, body silvery-grayish to brownish dorsally, becoming pale white on belly. Posterior end of adipose fin with an ovoid blackish spot.

Distribution. Known from India, Bangladesh, Pakistan, and Myanmar. In Arunachal Pradesh: see material examined.

Economic importance. This species is considered as a delicious food fish.

Conservational status: This fish is a widely spread species occurring in the multiple basins of Indian ocean. IUCN status: Least concern (LC).

Sperata seenghala **(Sykes 1839)**
Fig. 148.

FIG. 148 *Sperata seenghala*, RGUMF-174; 278 mm SL: lateral view.

Platystoma seenghala Sykes 1839, London part 6 for 1838: 164; type locality: Deccan, India.

Sperata seenghala: Kottelat, 2013: 266 (valid).

Materials examined. RGUMF-174, 7 exs., 196–382 mm SL, Subansiri River at Gerukamukh, Arunachal Pradesh (AP); RGUMF-unregistered, 04 exs., 132–235.6 mm SL, Ranga River at Kimin, Papum Pare district, AP.

Vernacular names in Arunachal Pradesh. *Ari* (in Nyishi, Adi, & Galo).

Diagnosis. Adipose-fin base length shorter than the dorsal-fin base length; adipose-fin origin almost opposite or slightly anterior to anal origin; snout truncate, 13–15 gill rakers on the first branchial arch; interneural shield rugose and ovoid.

Coloration. In live, body brownish-gray along back, and lighter on sides, pale silvery below lateral line and beneath. A rounded blackish spot on posterior end adipose-fin present.

Distribution. Known from India, Bangladesh, and Pakistan. In Arunachal Pradesh: see material examined.

Economic importance. This species is considered as a delicious food fish.

Conservational status: This fish is a widely spread species occurring in the Ganga-Brahmaputra and Indus river basins. IUCN status: Least concern (LC).

63. Genus: *Olyra* McClelland, 1842

Olyra McClelland 1842: 588. Type species: *Olyra longicaudatus* McClelland 1842. Type by subsequent designation.

Body elongated (eel like), flattened in pre-pelvic region and compressed behind; head short and depressed; barbels 4 pairs; gill opening wide, its functional part greatly restricted by flaps of skin along lower edges; rayed dorsal fin inserted above pelvic fin with 8 rays, without any spines; adipose fin long; pectoral fin with a sharp serrated spine; anal fin with more than 24 rays; and caudal fin forked, often with filamentously extended upper simple principle rays. The genus is also osteologically diagnosed by its exclusion of the prootic from the trigeminofacial foramen, supraoccipital spine greatly reduced and no ectopterygoid.

Key to the species of *Olyra*:

1a. Adipose-fin base longer than the dorsal-fin base length.................................... *O. longicaudata*

1b. Adipose-fin base shorter than the dorsal-fin base length .... *O. praestigiosa*

***Olyra longicaudata* McClelland 1842**

Fig. 149.

FIG. 149 *Olyra longicaudata*, RGUMF-394, 103 mm SL: lateral view.

Olyra longicaudata McClelland 1842, 2: 588, pl. 21, Fig. 1; type locality: Khasi hills, Meghalaya, India.

Olyra longicaudata: Kottelat, 2013: 264 (valid).

Materials examined. RGUMF-394, 24 exs., 33.0–103 mm SL, Dikrong River at Doimukh, Papum Pare district, Arunachal Pradesh (AP). RGUMF-unregistered, 10 exs., 63.4–82.0 mm SL, Ranga River at Kimin, Papum Pare district, AP; RGUMF-unregistered, 12 exs., 52.4–71.2 mm SL, Subansiri River in Lower Subansiri district, AP.

Vernacular name in Arunachal Pradesh. *Tayek* (in Nyishi & Galo), *Tayek/Bayek* (in Adi).

Diagnosis. Spine absent in dorsal fin. Pectoral spine serrated at both the anterior and posterior edges; pelvic fin not reaching half the length of distance between the pelvic and anal fins; adipose-fin base long with 29.0%–32.0% SL, post-adipose distance 7.0%–10% SL; and caudal-peduncle depth 5.0%–6.0% SL.

Coloration. In live, body brownish to grayish, superimposed with three blackish longitudinal bands: one midlateral band along the lateral line, one above and another below the midlateral stripe.

Distribution. Known only from the Brahmaputra River drainages in India. In Arunachal Pradesh: see material examined.

Economic importance. This species is an ornamental fish.

Conservational status: This fish is a widely spread species occurring in the Brahmaputra River basin only. IUCN status: Least concern (LC).

***Olyra praestigiosa* Ng & Ferraris 2016**

Fig. 150.

FIG. 150 *Olyra praestigiosa*, RGUMF-393, 101 mm SL: (A) lateral, and (B) dorsal views.

Olyra praestigiosa Ng & Ferraris, 2016: 382, Figs. 1 and 2, type locality: Chel River, Kalimpong Subdivision, Darjeeling district, West Bengal, India.

Materials examined. RGUMF-393, 10 exs., 64.2–101 mm SL: Dikrong River at Doimukh, Papum Pare district, Arunachal Pradesh (AP). RGUMF-unregistered, 05 exs., Ranga River at Kimin, Papum Pare district, AP; RGUMF-unregistered, 01 exs., Subansiri River in Lower Subansiri district, AP.

Vernacular names in Arunachal Pradesh. *Tayek* (in Nyishi, Galo), *Tayek/Bayek* (in Adi).

Diagnosis. Upper jaw projecting slightly beyond lower jaw; adipose-fin base shorter than the dorsal-fin base length; pectoral spine serrated on both edges with 7–12 serrae along the anterior and 8–15 retrorse serrae on the posterior edges, and eyes widely separated with interorbital distance 30%–37% HL.

Coloration. In live, body brownish on dorsal and lateral surfaces, fading to pale on ventral surfaces. Three dark brownish, moderately broad longitudinal stripes present on the flanks. Adipose fin, and fin rays and membranes of median fins dark brown.

Distribution. Known from the Brahmaputra River drainage in northeastern India and Bangladesh. In Arunachal Pradesh: see material examined.

Remarks. *O praestigiosa* differs from *O. longicaudata* in having a shorter adipose-fin base (9.9%–16.0% SL vs. 29.0–32.0). *A. kempi* is presently treated as synonym of *A. longicaudata*, however, original figure of the former species have very short adipose fin, thus required a thorough investigation.

Economic importance. This species is a potential ornamental fish.

Conservational status: This fish is a widely spread species occurring in the Ganga-Brahmaputra river basins. IUCN status: Not assessed (NA).

16. Family: Amblycipitidae

64. Genus: *Amblyceps* Blyth, 1858

Amblyceps Blyth 1858: 281; *Amblyceps caecutiens* Blyth 1858. Type by monotypy.

Body elongated; subcylindrical; abdomen rounded; head small broad, depressed; covered with thick skin; snout broadly rounded; mouth wide, anterior; eyes small superior; lips fleshy; nostrils close together, separated by nasal barbel; villiform teeth in bands on jaws, palate edentate; 4 pairs of barbels, 1 each of nasal and maxillary and 2 mandibular; a fold of skin above and anterior to pectoral-fin base, and immediately behind gill openings; dorsal fin inserted very far forwards, nearly above pectoral fin origin with a spine and 5-6 rays; adipose dorsal fin present; pectoral fin with a weak spine, lateral line present or absent. The genus is osteologically diagnosed as: epiphyseal commissure of supraorbital sensory canal is immediately located anterior to and not passing through epiphyseal bar; anterior cranial fontanel constricted abruptly near its posterior end to contribute bony support to epiphyseal commissure from frontal; fifth ceratobranchial expands medially at its posterior tip; pinnate like rays are present along anterior margin of procurrent caudal fin rays immediately anterior to upper and lower outermost branched caudal-fin rays; optic and trigeminal foramina in equal size; upper hypural fused with compound centrum.

Key to the species of *Amblyceps*:

1a. Caudal fin emerginate..2
1b. Caudal fin forked..3
2a. Large adult size reaching up to 160 mm SL, upper jaw longer than the lower jaw.. *A. apangi*
2b. Small adult size reaching up to 55.8 mm SL, upper jaw shorter than the lower jaw.. *A. laticeps*
3a. Pinnate like rays between median caudal-fin rays absent........................ ....*A. waikhomi*
3b. Pinnate like rays between median caudal-fin rays present..................................... 4
4a. Body depth at anus 10.1%–11.1% SL and total vertebrae 38 ....*A. arunachalensis*
4b. Body depth at anus 12.5%–15.8% SL and total vertebrae 34–36.............................. *A. mangois*

Amblyceps arunachalensis Nath & Dey 1989
Fig. 151.

FIG. 151 *Amblyceps arunachalensis*, RGUMF-117, 90.0 mm SL: lateral view.

Amblyceps arunachalensis Nath & Dey 1989: 3, 1, Fig. 3; type locality: Dikrong River, Arunachal Pradesh, India.

Amblyceps arunachalensis: Darshan et al., 2016: 8 (valid).

Materials examined. RGUMF-117; 4 exs., 45.0–100.0 mm SL, Subansiri River at Daporijo, Lower Subansiri district, Arunachal Pradesh (AP). RGUMF-unregistered, 8 exs., 47.3–94.6 mm SL, Dikrong at Doimukh, AP.

Vernacular name in Arunachal Pradesh. *Tayek* (in Nyishi & Galo), *Tayek*/*Bayek* (in Adi).

Diagnosis. Skin on head and body tuberculated; lower jaw slightly longer than upper; caudal fin fork, upper lobe longer than lower lobe (31.4%–32.5% SL vs. 22.0–22.8); presence of pinnate like rays between median caudal-fin rays; adipose-fin base length 18.1%–19.8% SL, total vertebrae 38; body slender with depth at anus 10.1%–11.1% SL, 12.5–15.8; dorsal fin with a spine and 6 rays; lateral line incomplete, extending to vertical through middle of dorsal-fin base.

Coloration. In 10% formalin, body dull grayish to dark brownish, fins with tinged grayish.

Distribution. Known only from the Brahmaputra basin of Assam and Arunachal Pradesh (see material examined section).

Remarks. *A. arunachalensis* differs from *A. mangois* in having a slender body (depth at anus: 10.1%–11.1% SL vs. 12.5–15.8) and more number of vertebrae (38 vs. 34–36).

Economic importance. This species is considered as an ornamental fish.

Conservational status: This fish is a widely spread species occurring in the Brahmaputra River basin only. IUCN status: Not assessed (NA).

***Amblyceps apangi* Nath and Dey 1989**

Fig. 152.

FIG. 152 *Amblyceps apangi*, RGUMF-114, 110.0mm SL: lateral view.

Amblyceps apangi Nath & Dey, 1989: 2, Fig. 2, type locality: Dikrong River, Arunachal Pradesh, India.

Amblyceps apangi: Darshan et al., 2016: 9 (valid).

Materials examined. RGUMF-118, 160.0mm SL, India, Arunachal Pradesh, Papum Pare district, Dikrong River at Sagalee, Brahmaputra basin, January 7, 2005. RGUMF-114–116; 15 exs., 71.5–120mm SL, Subansiri River at Daporijo; Panye River at Kalpangi, Papum Pare district Arunachal Pradesh (AP) RGUMF-unregistered, 6 exs., 47.9–116.8mm SL, Dikrong at Doimukh, Papum Pare district, AP.

Vernacular name in Arunachal Pradesh. *Tayek* (in Nyishi & Galo), *Tayek* or *Bayek* (in Adi).

Diagnosis. Skin on head and body smooth, jaws equal, no pinnate like rays between median caudal-fin rays; caudal fin lunate or truncate; adipose fin very long with length 24.3%–32% SL and not confluent with caudal fin; and lateral line complete.

Coloration. In live, dorsal and lateral surface of head and body pale yellow, ventrally lighter. Fins color similar to body color but distal margin hyaline.

Distribution. Known only from the Brahmaputra basin of Assam and Arunachal Pradesh.

Remarks. Maximum length of the species recorded from Arunachal Pradesh is ca. 160mm SL.

Economic importance. This species consider as an ornamental fish.

Conservational status: This fish is a widely spread species occurring in the Brahmaputra river basin only. IUCN status: Least concern (LC).

***Amblyceps laticeps* (McClelland 1842)**

Fig. 153.

FIG. 153 *Amblyceps laticeps*, RGUMF-379, 55.8 mm SL: dorsal view.

Olyra laticeps McClelland, 1842: 588, pl. 21, Fig. 2, type locality: Kasyah Mts [Khasi Hills] India.

Amblyceps laticeps: Darshan et al., 2016: 9 (valid).

Materials examined. RGUMF-379, 44.5–55.8 mm SL, Kameng River at Tippi, West Kameng district, Arunachal Pradesh (AP). RGUMF-380, 45.1–52.6 mm SL, Dibang River, Lower Dibang Valley district, AP.

Vernacular names in Arunachal Pradesh. *Tayek* (in Nyishi & Galo), *Tayek* or *Bayek* (in Adi).

Diagnosis. Upper jaw short, lower jaw strongly projected anteriorly and distinctly exposed, caudal fin emerginated, 41–43 vertebrae; adult size very small, mature at 44.5 mm SL.

Coloration. In 10% formalin, body brownish-gray to dark-gray, fins with tinged pale grayish.

Distribution. Known from the Ganga-Brahmaputra river system of India. In Arunachal Pradesh: see material examined.

Economic importance. This species consider as an ornamental fish.

Conservational status: This fish is a widely spread species occurring in the Ganga-Brahmaputra river basins. IUCN status: Least concern (LC).

***Amblyceps waikhomi* Darshan et al., 2016**

Fig. 154.

FIG. 154 *Amblyceps waikhomi*, RGUMF-271, paratype, 30.4 mm SL: lateral view.

Amblyceps waikhomi Darshan et al., 2016: 2, Fig. 1, 4c; type locality: stream at Nongkon village draining into the Noa Dehing River, Brahmaputra River basin, Namsai District, Arunachal Pradesh, India (valid).

Materials examined. ZSI/APRC/P-1125, 01 ex., holotype, 42.9 mm SL, Nongkon stream at Nongkon village draining into Noa Dehing River (Brahmaputra basin), 27°36′05″N, 95°50′51″E in Namsai District, Arunachal Pradesh, India. RGUMF-269, 01 ex., 40.2 mm SL, collection data same as above; RGUMF-270, 01 ex., 37.4 mm SL, Nongkon stream draining into Noa-Dehing River (Brahmaputra basin), Namsai District, AP. RGUMF 271, 30.4–44.7 mm SL, 2 exs., Nongkon stream draining into Noa-Dehing River (Brahmaputra basin), Namsai District, AP, December 6, 2014.

Vernacular names in Arunachal Pradesh. *Tayek* (in Nyishi & Galo), *Tayek/Bayek* (in Adi).

Diagnosis. Body robust, lower jaw slightly longer than upper; body depth at anus 17.0%–20.3% SL, and total vertebrae 34–35; no pinnate like rays between median caudal-fin rays; head large (length: 22.1%–24.3% SL; width: 20.0%–21.9% SL; depth: 13.9%–18.0% SL). Presence of pinnate like rays on outer margin of principal ray of caudal fin. Lateral line incomplete, terminating at a point slightly anterior to vertical through dorsal fin origin. First branchial arch with 2+7 gill rakers. Caudal fin deeply forked with i,7,8,i principal rays and upper lobe longer.

Coloration. In 10% formalin, body and head brownish on dorsal and lateral surfaces, and ventrally creamy.

Distribution. Currently known only from the Noa-Dehing River (Brahmaputra basin), Namsai District, Arunachal Pradesh.

Economic importance. This species consider as an ornamental fish.

Conservational status: The species is restricted in the Noa-Dehing River only (Brahmaputra basin) in the state Arunachal Pradesh. IUCN status: Not assessed (NA).

Amblyceps mangois (Hamilton 1822)

Fig. 155.

FIG. 155 *Amblyceps mangois*, RGUMF-429, 25.7 mm SL: lateral view.

Pimelodus mangois Hamilton, 1822: 199, 379; type locality: Tanks of North Behar, Kosi River, India.

Amblyceps mangois: Darshan et al., 2016: 8 (valid).

Material examined. RGUMF-429, 1 ex., 25.7 mm SL, Dikrong River at Doimukh, Papum Pare district, Arunachal Pradesh.

Vernacular name in Arunachal Pradesh. *Tayek* (in Nyishi & Galo), *Tayek/Bayek* (in Adi).

Diagnosis. Both jaws equal; caudal fin deeply forked, upper lobe slightly longer than the lower lobe, pinnate like rays present between the median caudal-fin rays; adipose-fin base length 16.9–25.2 mm SL, body depth at anus 12.5–15.8 mm SL; head small (length 18.8%–21.3% SL; width 17.1%–18.3% SL, depth 11.6%–13.2% SL); body depth at anus 12.5–15.8.

Coloration. In live, body olive brownish above and lighter beneath.

Distribution. Widely distributed in the Ganga-Brahmaputra basin.

Economic importance. This fish is considered as an ornamental fish.

Conservational status: This fish is a widely spread species occurring in the Ganga-Brahmaputra river basins. IUCN status: Least concern (LC).

17. Family: Sisoridae

Subfamily: Sisorinae

65. Genus: *Sisor* Hamilton, 1822

Sisor Hamilton 1822: 208, 379. Type species: Masc. *Sisor rabdophorus* Hamilton 1822. Type by monotypy.

Body very elongated, depressed anteriorly; head moderately narrow; skin with rough ridges; mouth small, inferior, upper jaw longer; lips thick and fleshy; post-labila groove on lower jaw prominent; gill opening narrow, membranes joined in broad isthmus; four branchiostegal rays; teeth present or absent on jaws; eyes small; barbels six pairs: one of maxillary and five of mandibular; rayed dorsal with a weakly anteriorly serrated spine; pectoral fin spine serrated on both edges; caudal fin truncate, its first upper simple principal ray produced into a long whip like filament; lateral line with a series of small rough bony plates.

***Sisor rhabdophorus* Hamilton 1822**

Fig. 156.

FIG. 156 *Sisor rhabdophorus*, RGUMF-377, 150.6 mm SL: dorsal view.

Sisor rhabdophorus Hamilton, 1822: 208, 379; type locality: Bhagirathi River at crossing point between Kalna (Bardhaman district) and Nrisinghapur [Nadia district], West Bengal, India.

Sisor rhabdophorus: Shrestha, 2008: 192 (valid).

Material examined. RGUMF- 377, 01 ex., 150.6 mm SL, Tengapani River at Namsai, Namsai district, Arunachal Pradesh.

Vernacular names in Arunachal Pradesh. Not available (in Nyishi, Adi, & Galo).

Diagnosis. Bony plates on surface of body; first principal caudal-fin ray enlarged and produced into a long whip-like filament; body depth at anus 5.0%–5.4% SL; lateral line ossicles 66–70; pectoral spine serrated on both edges with 27–29 serrae anteriorly and 6–15 serrae on posterior margin.

Coloration. In live, body blackish above and ventrally whitish to silvery.

Distribution. Known from the Ganges-Brahmaputra river drainage of India; Nepal and Bangladesh. In Arunachal Pradesh: abundantly occurring in Namsai district.

Economic importance. This species is considered as an ornamental fish.

Conservational status: This fish is a widely spread species occurring in the Ganga-Brahmaputra river basins. IUCN status: Not assessed (NA).

66. Genus: *Bagarius* Bleeker, 1853

Bagarius Bleeker 1853: 121. Genus: *Pimelodus buchanani* Bleeker 1853. Type by monotypy.

Body elongated and depressed up to pelvic insertion; head broad, moderately depressed and large, covered with keratinized rugose skin; eyes small, subcutaneous, placed dorsally, and located at posterior half of head; mouth wide, terminal or slightly inferior; barbels 4 pairs: 1 each of nasal and maxillary, and 2 of mandibular; gill opening wide, continued forward ventrally, branchiostegal membranes free from isthmus and also from each other but overlapping; branchiostegal rays 9–13; lips thick; premaxillary bone on each with two ossifications; heterodont dentition in lower jaw, consisting of 2–3 outer rows of closely set conical teeth and 1–2 inner rows of widely separated much larger conical teeth; palate edentate; caudal peduncles small and elongated; dorsal fin inserted above the base of pectoral fin; with a strong smooth spine and 7 branched rays, outermost element may bears a filamentous extension; adipose dorsal fin long; pectoral fin with a spine serrated finely on posterior margin, followed by 9–14 rays, outer rays with a filamentous extension; anal fin with 13–17 rays; pelvic fin with 6 rays; caudal fin deeply forked, both lobes with filamentous extensions.

Bagarius bagarius (Hamilton 1822)
Fig. 157.

FIG. 157 *Bagarius bagarius*, RGUMF-427, 253.9 mm SL: lateral view.

Pimelodus bagarius Hamilton, 1822: 186, 379, pl. 7, Fig. 62; type locality: Ganges and its tributaries, India.

Bagarius bagarius: Kottelat, 2013: 221 (valid).

Materials examined. RGUMF-120; 1 ex., 420 mm SL, Dirang River in West Kameng District, Arunachal Pradesh (AP). RGUMF-427, 01 ex., 253.9 mm SL, Tengapani at Namsai, Namsai district, AP.

Vernacular name in Arunachal Pradesh. *Mein-lomen/Ngone* (in Nyishi & Adi).

Diagnosis. Pelvic-fin origin anterior to a vertical line through base of last dorsal-fin ray; adipose origin at the level behind anal fin origin; neural spines 4–8 distally expanded; body depth 7.6%–8.1% SL; eye diameter 10.9%–11.4% HL; gill rakers on the first gill arch 6–9; pectoral fin with 9–12 branched rays and anal fin with 13–14 branched rays.

Coloration. In live, three blackish pigmented bands or blotches on the body: one extending from dorsal-fin base to pelvic fin, second from adipose-fin base to anal fin, and last on posterior part of caudal peduncle. Body dark yellow on dorsal and lateral surfaces of head and body, light yellow on ventral region. A few small brown spots irregularly distributed throughout lighter color parts of body. All fins light yellow, with irregularly distributed small brown spots.

Distribution. Known from the Ganges and Brahmaputra drainages in India, Bangladesh, Bhutan, and Nepal. In Arunachal Pradesh: abundantly occurring in Namsai and Lohit district.

Remarks. *B. bagarius* differs from *Bagarius yarrelli* in having pelvic-fin origin usually anterior (vs. posterior) to a vertical line through base of last dorsal-fin ray; adipose-fin origin slightly too markedly posterior (vs. anterior to, on, or slightly posterior) to a vertical line through anal-fin origin; total external gill rakers on first gill arch 6–9 (vs. 8–11).

Economic importance. This species is a highly prized food fish.

Conservational status: This fish is a widely spread species occurring in the Gang-Brahmaputra river basins. IUCN status: Near threatened (NT).

67. Genus: *Gagata* Bleeker, 1858

Gagata Bleeker 1858: 204, 206. Type species: Masc. *Pimelodus gagata* Hamilton 1822. Type by absolute tautonymy.

Body compressed; head short with a distinct median groove above; mouth ventral, narrow, and transverse; lips thick; fimbriated; villiform teeth on jaws, no teeth on palate; barbels 4 pairs; 1 pair maxillary, 2 pair mandibulars, inserted on a transverse row at same level; 1 pair nasal barbel, its broad base separated the two closely set nostrils; gill membranes broadly joined with isthmus; rayed dorsal fin with a smooth spine; inserted above the middle of pectoral fin; pectoral fin with a spine (serrated internally) and 8–9 branched rays; lateral line complete; caudal fin deeply forked.

Gagata cenia (Hamilton 1822)

Fig. 158.

FIG. 158 *Gagata cenia*, RGUMF-unregistered, 47.2 mm SL: lateral view.

Pimelodus cenia Hamilton, 1822: 174, pl. 31, Fig. 57; type locality: Rivers of North Bengal, India.

Gagata cenia: Gurung et al., 2013: 4884 (valid).

Materials examined. RGUMF-0121, 4 exs., 43.0–55.0 mm SL, Poma River at Ramgath, Papum Pare district, Arunachal Pradesh. RGUMF-unregistered, 01 ex., 47.2 mm SL, Lohit River at Alubari Ghat, Lohit district, Arunachal Pradesh.

Vernacular name in Arunachal Pradesh. *Narang* (in Adi).

Diagnosis. A median longitudinal groove on the head extending only up to the base of occipital process; snout longer than eye diameter; body with five dark saddles extending ventrally only to the lateral line, middle of the upper and lower lobes of caudal fin with one black blotch each; dorsal fin with black spot on distal part of anterior rays.

Coloration. In live, body ground color creamish, overlay with five grayish saddles extending to lateral line. Ventral safari of body whitish to silvery. Distal parts of dorsal, anal, pectoral, pelvic, and adipose-fin with no black marking.

Distribution. Known from the Ganges-Brahmaputra system in India, Bangladesh, Nepal, and Pakistan. In Arunachal Pradesh: see material examined.

Economic importance. This species have high ornamental value.

Conservational status: This fish is a widely spread species occurring in the Ganga-Brahmaputra and Indus river basins. IUCN status: Least concern (LC).

68. Genus: *Gogangra* Roberts, 2001

Gogangra Roberts 2001: 83. Type species: *Pimelodus viridescens* Hamilton 1822. Type by being a replacement name.

Head short and broad; mouth wide and inferior, jaws with conical teeth; palate edentulous; four pair of barbels, one pair each of nasal and maxillary, two pairs on mandibular, maxillary barbel without a conspicuous membrane posteriorly, the outer and inner mandibular barbels widely separated with the origin of inner barbels anterior to origin of outer barbels, the branchiostegal membranes free from isthmus; Y-shaped mesethmoid, not greatly expanded anteriorly.

***Gogangra viridescens* (Hamilton 1822)**

Fig. 159.

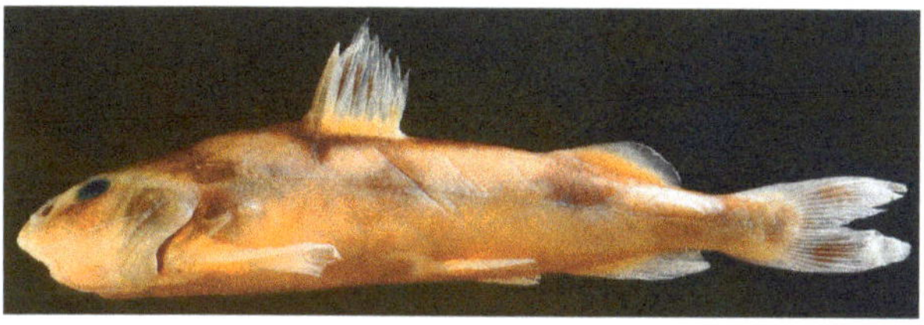

FIG. 159 *Gogangra viridescens*, RGUMF-122, 56.0mm SL: lateral view.

Pimelodus viridescens (Hamilton), 1822: 173, 377, pl. 11, Fig. 56; type locality: Rivers of North Bengal, India.

Gogangra viridescens: Ferraris, 2007: 395 (valid).

Material examined. RGUMF-122, 2 exs., 54.0–56.0mm SL, Dikrong River at Doimukh, Papum Pare district, Arunachal Pradesh.

Vernacular name in Arunachal Pradesh. *Narang* (in Adi).

Diagnosis. Inner and outer mental barbel widely separated; inner mental barbel inserted slightly anterior to origin of outer mental barbel; maxillary barbels much shorter than head; anterior and posterior fontanel broad and the two are separated by a narrow bridge, lateral fontanel absent, pectoral girdle provided with rugose coracoid process.

Coloration. In live, body brownish to pale grayish with a bright viridescent horizontal silvery mark in the membrane connecting the opercle to the side of the cranium. Flank with four broad bands above lateral line, Dorsal and caudal fins with a blackish band.

Distribution. Known from the Ganga-Brahmaputra river drainages in India, Bangladesh, and Nepal. In Arunachal Pradesh: see material examined section.

Remarks. *Gogangra viridescens* was previously considered as a species of *Nangra* Day (1877). Members of the *Gogangra* are distinguished from that of *Nangra* in having a Y-shaped mesethmoid that is not greatly expanded (vs. fan-shaped mesethmoid that is greatly expanded), and the maxillary barbel without a conspicuous membrane posteriorly (vs. with conspicuous membrane) (Roberts and Ferraris, 1998).

Economic importance. This species have high ornamental value.

Conservational status: This fish is a widely spread species occurring throughout the Ganga-Brahmaputra basin. IUCN status: Least concern (LC).

Subfamily: Glyptosterninae

69. Genus: *Creteuchiloglanis* Zhou et al. 2011

Creteuchiloglanis Zhou, Li & Thomson 2011: 227. Type species: *Creteuchiloglanis longipectoralis* Zhou, Li & Thomson 2011. Type by original designation.

Post-labial fold interrupted, isthmus very broad, gill opening not extended to underneath the head, 14–16 branched pectoral-fin rays, lower lip not connected to the base of the maxillary barbel by a skin flap, sulcus present between the lip and the barbel; dentition homodont, teeth pointed, coniform in both jaws, premaxillary tooth patches in contact to each other but not confluent, not extended posterolaterally, and laterally blunt.

Key to species of *Creteuchiloglanis*:

1a. Posterior margin of pectoral fin distinctly reaching the pelvic-fin origin, predorsal length with 34.5%–37.5% SL... ....*C. kamengensis*

1b. Posterior margin of pectoral fin not reaching pelvic-fin origin, predorsal length shorter than 34.0% SL..2

2a. Milky-white patches on the inner posterior margin of skin flap connecting maxillary barbel and cheek; and whitish spot at the base of last pectoral-fin ray...*C. payjab*

2b. No marking on the inner posterior margin of skin flap connecting maxillary barbel and check, no distinct marking at the base of last pectoral-fin ray 3

3a Pre-pelvic length 45.4% SL, adipose fin base length 25.9% SL snout length 54.8% SL... *C. arunachalensis*

3b. Pre-pelvic length 39.4–39.5, adipose fin base length 32.4–34.3 snout length 58.1%–60.1% SL.. *Creteuchiloglanis* sp.

***Creteuchiloglanis kamengensis* (Jayaram 1966)**
Fig. 160.

FIG. 160 *Creteuchiloglanis kamengensis*, RGUMF-441, 108 mm SL: (A) dorsal, and (B) ventral views.

Euchiloglanis kamengensis Jayaram, 1966: 85, Fig. 1; type locality: India: Norgum River at Kalaktang, Kameng frontier Division, Nefa, Arunachal Pradesh, India.

Creteuchiloglanis kamengensis: Sinha & Tamang, 2014: 5 (valid).

Materials examined. RGUMF-128, 1 ex., 75.0 mm SL, Baichom River at Rupa, Brahmaputra basin, Arunachal Pradesh (AP). RGUMF 441, 1 ex., 108 mm SL, Kameng River at Sergoan, West-Kameng district, AP, collected by Achom Darshan, M. Stalin Singh & Mr. Dulan, November 29, 2014. RGUMF-448, 1 ex., 114 mm SL, Kameng River at Rupa, West Kameng district, AP, Leki Wangchu, March 2015.

Vernacular names in Arunachal Pradesh. *Niipe* (in Apatani); *Charba* (in Nyishi), *Kesari* (in Adi), *Batpala* (in Monpa).

Diagnosis. Adpressed tip of pectoral-fin just reaching pelvic-fin origin; absence of pale patches on body except on occipital region; caudal-fin slightly obliquely truncated, lower portion slightly longer; lateral line complete; pectoral fin with i,16 rays; pelvic fin with i,5 rays and anal fin with i,6 rays.

Coloration. In live, body yellowish to brownish-gray dorsally, ventral surface of head, belly and abdominal surfaces dull whitish. Anterior and posterior region of adipose fin light orange; origin and the base of last branched rays of pectoral and pelvic fin with pale spots.

Distribution. Known only from the Norgum and Kameng rivers in West Kameng district, Arunachal Pradesh. Reports of this species from the upper reaches of Lohit River in China need confirmation.

Remarks. *C. kamengensis* differs from *C. arunachalensis* by having longer adipose-fin base (32.7–37.3 vs. 25.9% SL) and pectoral fin (25.0–28.7 vs. 23.1% SL).

Economic importance. This species is considered as an ornamental fish.

Conservational status: This fish was previously assumed to be a widespread species in the Brahmaputra River basin. However, we observed that the species is restricted in Norgum and Kameng River drainages only. Recently, *C. payjab* and *C. arunachalensis* are also been described from its neighboring drainage: the Siang and Ranga river, respectively. A member of this genus, encountered from the Tawang river (Manas drainage of Brahmaputra basin) is also a distinct undescribed species. IUCN status: Not assessed (NA).

***Creteuchiloglanis payjab* Darshan et al. 2014**

Fig. 161.

FIG. 161 *Creteuchiloglanis payjab*, RGUMF-249, holotype, 172 mm SL: (A) lateral, and (B) ventral views.

Creteuchiloglanis payjab Darshan et al., 2014: 74, Figs. 1 and 2; type locality: Arunachal Pradesh state, West Siang district, Yomgo River at Mechuka, a tributary of Siang River, Brahmaputra River basin, India (valid).

Materials examined. RGUMF-249, 1 ex., 172 mm SL, Yomgo River at Mechuka, a tributary of Siang River, Brahmaputra River basin, Arunachal Pradesh, India. RGUMF-251–254, 4 exs., 110.2–155.1 mm SL; ZSI/V/APRC/P-921, 1 ex., 153.5 mm SL (same data).

Vernacular names in Arunachal Pradesh. *Nyabe* (in Memba), *Payjab* (in Bokar, Ramo and Libo of Mechukha); *Niipe* (in Apatani); *Charba* (in Nyishi); *Kesari* (in Adi); *Batpala* (in Monpa).

Diagnosis. Milky-white patches at the base of last pectoral-fin ray and also on the inner posterior margin of skin flap connecting maxillary barbel; head and abdominal region strongly depressed and broad, then laterally compressed posteriorly. Dorsal fin without spine, with i,6 rays; pectoral fin without spine with i,14–15 rays; pelvic fin with i,5 rays; anal fin with i,3–4 rays and caudal fin truncate with i,15,i rays. Adipose fin long, anterior margin rising gently and then almost straight for the entire length.

Coloration. In live, dorsal and lateral surface of head and body brownish gray, ventral surface of head and abdomianal region dirty white, remaining portions brownish, milky-white patches at the base of last pectoral-fin ray and on inner posterior margin of skin flap connecting maxillary barbel. Basal portion of first ray and last ray of the pelvic-fin pale yellow, distal portion of adipose fin pale yellow. Dorsal and ventral margin of caudal fin yellowish and distal tip of pair fins yellowish with reddish patches in pectoral fin.

Distribution. Known only from the Mechuka, Manigong, Pidi, and Tato villages along stretches of the Yomgo River (= Siyom River), a headwater of the Siang River, Brahmaputra river basin, Arunachal Pradesh, India.

Economic importance. This species consider as an ornamental fish.

Conservational status: The known distribution of this species is restricted to the fast flowing coldwater (high altitude) streams of the upper stretches of Yomgo River. Based on the information of local villagers, the abundance of this species has been drastically reduced after introducing an exotic trout, *Oncorynchus mykiss*. Removal of this invasive exotic species from the natural habitat of *C. payjab* may be an ideal conservational step. IUCN status: Not assessed (NA).

***Creteuchiloglanis arunachalensis* Sinha & Tamang 2014**
Fig. 162.

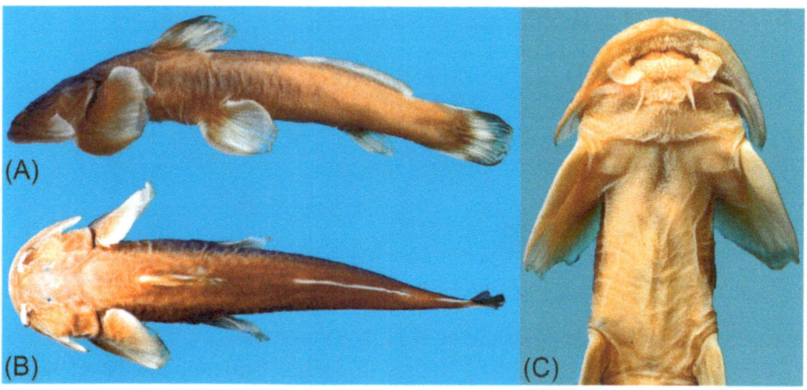

FIG. 162 *Creteuchiloglanis arunachalensis*, ZSI/V/APRC/P-844, holotype, 87.1 mm SL: (A) lateral, (B) dorsal, and (C) ventral part of mouth and belly.

Creteuchiloglanis arunachalensis Sinha & Tamang, 2014: 2, Figs. 1, 2a; type locality: Pange River at Aro-Lenching, a tributary of Brahmaputra River, Ziro, Lower Subansiri District, Arunachal Pradesh, northeastern India (valid).

Material examined. ZSI/V/APRC/P-844, 87.1 mm SL, Pange River at Aro-Lenching, a tributary of Brahmaputra River, Ziro, Lower Subansiri District, Arunachal Pradesh, northeastern India, Coll. B. Sinha et al., June 9, 2013.

Vernacular names in Arunachal Pradesh. *Niipe* (in Apatani); *Ngiipe (in Apatani); Charba* (in Nyishi), Kesari (in Adi), *Batpala* (in Monpa).

Diagnosis. Body surface with pale patches except on occipital region; no whitish patch on the basal posterior corner of the maxillary barbel; pectoral fin short, its distal margin not reaching the pelvic-fin origin; pelvic fin not reaching at the vertical through origin of adipose fin; post-adipose distance (10.1% SL); obtusely rounded distal margin rectal lobe; caudal fin with a broad bean-shaped medial band, its posterior margin slightly convex.

Coloration. In 10% formalin, over all the body light brownish except ventral portion anterior to the anal origin light creamy. Pale patches except on occipital region, scattered over the brownish background color. Anterior margin of dorsal fin grayish, distally dusky and hyaline. Caudal fin with a brownish base and a distal black band.

Distribution. Known only from the Pange River (Ranga drainage), Lower Subansiri district, Arunachal Pradesh, northeastern India.

Economic importance. This species is considered as an ornamental fish.

Conservational status: The species is restricted in the Pange River (upper reaches of the Subansiri drainage) only, a tributary of Brahmaputra basin. IUCN status: Not assessed (NA).

***Creteuchiloglanis* sp.**

Fig. 163.

FIG. 163 *Creteuchiloglanis* sp. RGUMF-450, 166.6mm SL: (A) lateral, and (B) ventral views.

Creteuchiloglanis sp. Darshan et al., Fig. 1; type locality, Tawangchu River at Granger village (headwater of Manas drainage, Brahmaputra river basin), Tawang district, Arunachal Pradesh state, India, 27°33′39.9″N, 91°53′15.7″E, elevation 1946m (in press).

Materials examined. RGUMF 450, 1, 152mm SL, India: Arunachal Pradesh state, Tawang district, Tawangchu River at Granger village, headwater of Manas drainage, Brahmaputra river basin, 27°33′39.9″N, 91°53′15.7″E, elevation 1946m, Leki Wangchu, March 8, 2015. RGUMF 449, 138.7mm SL same data as above.

Vernacular names in Arunachal Pradesh. *Niipe* (in Apatani); *Ngiipe* (in Apatani); *Charba* (in Nyishi), Kesari (in Adi), *Batpala* (in Monpa).

Diagnosis. The new species differs from all congeners in having a unique combination of the following characters: pectoral fin not extending to the origin of pelvic fin and very broad with stretched transverse span 77.3%–80.8% Pt-Pl or 21.0%–22.0% SL; distance between the origins of pectoral and pelvic with 26.9%–27.5% SL; moderately long snout with 58.1%–63.9% HL; eye diameter 7.9%–9.1% HL; first branchial arch with 2 + 8 = 10 gill rakers; caudal peduncle depth with 6.9–7.2; and post-adipose distance 9%–10.8% SL.

Coloration. In 10% formalin, overall body color brownish to grayish, pale creamish on ventral surfaces of head and belly. Pair fins origins and bases of last fin rays with pale spots. Distal margin of all fins except caudal fin with pale creamish margins. Basal half of caudal fin pale creamish, distally brownish.

Distribution. Known only from the Tawangchu River (Manas drainage), Tawang district, Arunachal Pradesh, northeastern India.

Economic importance. This species is considered as an ornamental fish.

Conservational status: The species is restricted in the Tawangchu River (Manas drainage). IUCN status: Not assessed (NA).

70. Genus: *Exostoma* Blyth, 1860

Exostoma Blyth 1860: 155; Neut. Type species: *Exostoma berdmorei* Blyth 1860. Type by subsequent designation.

Body elongated, depressed anteriorly up to pelvic fin base; nostrils close together and separated by a short flap-like nasal barbel; eyes minute and subcutaneous; mouth ventral and crescentric; lips thick and fleshy, papillated, continuous around mouth to form a sucker; teeth large and oar shaped, upper jaw teeth separated in two distinct patches; no teeth on palate; barbels four pairs; gill membranes confluent with isthmus; thorax without adhesive apparatus; dorsal fin inserted close to pelvic-fin origin, fin spine absent; adipose dorsal low, long; may continuous with caudal fin; pectoral fin short, not overlapping with pelvic fin, its first element unosseous and ventrally pleated; pelvic fin with six rays, its first element pleated ventrally, pectoral fin without a spine; lateral line complete.

Key to the species of *Exostoma*:

1a. Thick caudal peduncle with depth 8.7%–9.9% SL........................ *E. labiatum*

1b. Thin caudal peduncle with depth 3.6%–4.7% SL.............................. *E. tenuicaudata*

Exostoma labiatum **(McClelland 1842)**
Fig. 164.

FIG. 164 *Exostoma labiatum*, RGUMF-442, 45.3 mm SL: lateral view.

Glyptosternon labiatus McClelland, 1842: 588; type locality: Mishmee Hills [Arunachal Pradesh], India.

Exostoma labiatum: Zhang et al., 2016: 172 (valid).

Materials examined. RGUMF-439, 2 exs., 53.8–76.1 mm SL, Tezu River at Tezu, Lohit district, Arunachal Pradesh. RGUMF-442, 8 exs., 45.3–66.5 mm SL, Kurung River (Subansiri drainage), Kurung Kumey district, AP.

Vernacular name in Arunachal Pradesh. *Niipe* (in Apatani); *Herpakbellap* (in Nyishi), *Ngoge* (in Adi).

Diagnosis. Adipose-fin base length 26.6%–30.5% SL, adipose fin not confluent with caudal fin; caudal peduncle depth with 8.7%–9.9% SL; pectoral fin with 11–12 branched rays; caudal fin emerginated with i,8 + 7,i rays.

Coloration. In 10% formalin, over all the body uniformly light yellowish to grayish, ventrally creamish. Dorsal-fin rays light brownish gray and distal margin almost hyaline. Adipose fin with light grayish brownish with creamy distal margin. Anal-fin base and subdistal margin light grayish brown and tip hyaline. Caudal fin dark to light grayish brown.

Distribution. Known from India and China. In Arunachal Pradesh: see material examined.

Remarks. Mishmi hills (= Mishmee mountains of McClelland 1842), the type locality of *Exostoma labiatum*, is located at the junction of Lower Dibang Valley and Lohit districts of Arunachal Pradesh. However, many authors erroneously regarded the Mishmi hills as a part of the northeast Indian state Meghalaya.

Economic importance. This species consider as an ornamental fish.

Conservational status: The species is assumed to be restricted to the Dibang and Lohit Rivers in the Brahmaputra basin. IUCN status: Least concern (LC)

***Exostoma tenuicaudata* Tamang, Sinha & Gurumayum 2015**

Fig. 165.

FIG. 165 *Exostoma tenuicaudata*, ZSI/APFS/P-929, holotype, 108 mm SL: (A) lateral, (B) dorsal views, and (C) ventral view mouth part and belly.

Exostoma tenuicaudata Tamang et al., 2015: 441, Figs. 1–3, 5, 6A; type locality: Siang River, from Bomdo village on main road to Tuting, Upper Siang district, Arunachal Pradesh, India (valid).

Material examined. ZSI/APFS/P-929, holotype, 108.0mm SL; Upper Siang district, a fast flowing mountain drainage, confluence directly to Siang River, Arunachal Pradesh, India, about 3km from Bomdo village on main road to Tuting, 28°44.04′N 94°51.97′E, 429m asl; L. Tamang, October 26, 2011. ZSI/APFS/P-930, 4, 82.7–103.2mm SL; same data as holotype, October 26–28, 2011.

Vernacular names in Arunachal Pradesh. *Niipe* (in Apatani); *Herpak-bellap* (in Nyishi), *Ngoge* (in Adi).

Diagnosis. Adipose-fin base length slightly greater than three times the length of dorsal-fin base or its base length ranging from 30.8% to 32.9% SL, posterior end of adipose-fin not contiguous with upper caudal-fin procurrent rays; snout length 60.2%–61.8% HL; narrow caudal peduncle with depth 3.6%–4.7% SL; pectoral fin with i,11–12 rays; caudal fin weakly forked with i,7,7,i rays.

Coloration. In live, body brownish gray on dorsal and lateral surfaces, belly creamish yellow. Dorsal-fin base dark brown, its interradial membranes and distal margin hyaline. Adipose fin light grayish, anal fin light brownish with hyaline distal margin. Caudal fin blackish to light brownish.

Distribution. Known only from the Siang River, Upper Siang district, Arunachal Pradesh, India.

Economic importance. This species is considered as an ornamental fish.

Conservational status: The species is assumed to be restricted in the Siang River drainage only, a tributary of Brahmaputra basin. IUCN status: Not assessed (NA).

71. Genus: *Glyptothorax* Blyth, 1860

Glyptothorax Blyth 1860: 154. Type species: Masc. *Glyptosternon striatus* McClelland 1842. Type by subsequent designation.

Body elongated and depressed; skin smooth or covered with granules or tubercles; head depressed; mouth inferior, transverse and narrow; thorax with adhesive apparatus forms by longitudinal folds of skin with or without a pit or central depression; rayed dorsal fin with a smooth spine; adipose dorsal short; pectoral fin with posteriorly serrated spine; paired fins horizontally inserted broadly with first element with or without pleats below; caudal fin deeply forked; lateral line complete.

Key to species of *Glyptothorax* (Brahmaputra basin in Arunachal Pradesh):

1a. Ventral surfaces of both the pectoral-fin spine and the first simple ray of pelvic fin plicated................2

1b. Ventral surfaces of the pectoral-fin spine and first simple ray of pelvic fin lacking pleats......3

2a. A wedge-shaped central depression in the thoracic adhesive apparatus which open caudally, depressed region of the apparatus devoid of skin ridges... *G. striatus*

2b. The absence of the central depression on the thoracic adhesive apparatus...*G. pantherinus*
3a. Chevron-shaped thoracic adhesive apparatus....................................4
3b. Elliptical or leaf or rhomboidal-shaped thoracic adhesive apparatus..................................5
4a. Distance between the posterior end of pectoral-fin base and pelvic-fin origin is much greater than that of the distance between origins of pelvic and anal fin...................... *G. indicus*
4b. Distance between the posterior end of pectoral-fin base and pelvic-fin origin is equal to the distance between pelvic-fin origin and anal-fin origin.. ....*G. dikrongensis*
5a. Thoracic adhesive apparatus without depression.................................... 6
5b. Thoracic adhesive apparatus with depression.................................... 7
6a. No dark saddles on the back.. .... *G. telchitta*
6b. Three narrow dark brown saddles along the back .... *G. botius*
7a. Depression on thoracic adhesive apparatus in the form of a deep central pit, no ridges inside the pit..*G. cavia*
7b. Depression on thoracic adhesive apparatus in the form of longitudinal spindle, distinct ridges inside the depression..8
8a. Head length 21.4%–22.6% SL, snout length 52.9%–58.6% HL.............. .... *G. mibangi*
8b. Head length 22.7%–24.8% SL, snout length 46.6%–50.7% HL.............. ....*G. pashighatensis*

***Glyptothorax cavia* (Hamilton 1822)**
Fig. 166.

FIG. 166 *Glyptothorax cavia*, RGUMF-411, 217 mm SL: (A) lateral view, and (B) ventral view of thoracic adhesive apparatus and belly.

Pimelodus cavia Hamilton, 1822: 188, 378; type locality: rivers of Northern Bengal, India.

Glyptothorax cavia: Darshan et al., 2015: 120 (valid).

Materials examined. RGUMF-123, 2 exs., 100–170 mm SL, Noadihing River at Namdapha, Tirap district, Arunachal Pradesh (AP). RGUMF-411, 1 ex., 217 mm SL, Dibang River at Roing, Lower Dibang Valley, AP. RGUMF-unregistered, 2 exs., 86.2–102 mm SL, Dikrong River at Doimukh, Papum Pare district, AP.

Vernacular name in Arunachal Pradesh. *Ngurik* (in Nyishi), *Ngoling* (in Adi).

Diagnosis. Thoracic adhesive apparatus with a deep pit (Fig. 166B), surrounded by double diverging series of pleats; gill rakers on the first branchial arch with 1 on upper and 5 on lower arms (total six); upper jaw tooth plate very large, roofing anterior portion of mouth cavity; head length 27.6%–29.9% SL; predorsal distance 36.8%–41.3% SL; pre-pectoral length 22.0%–25.4% SL; preanal length 68.2%–70.5% SL; post-adipose distance (20.2–21.8 vs. 16.4%–17.8% SL) and slender caudal peduncle 4.8%–6.4% SL.

Coloration. In 10% formalin, body surface brownish with scattered dark brown spots. Dorsal and anal fins with two dark stripes: one at base, one submarginal separated by a white one in middle, fin edges white. Adipose, Pectoral and pelvic fins brown with white edges. Caudal fin with a black submarginal band.

Distribution. Known from the Ganga-Brahmaputra river drainage of India. In Arunachal Pradesh: see material examined.

Remarks. Largest size of this species recorded from Arunachal Pradesh is 217 mm.

Economic importance. This species is considered as an ornamental fish.

Conservational status: The species is widely distributed throughout the Ganga-Brahmaputra basin only. IUCN status: Least concern (LC).

Glyptothorax telchitta **(Hamilton 1822)**

Fig. 167.

FIG. 167 *Glyptothorax telchitta*, RGUMF-183, 68.9 mm SL: (A) lateral, and (B) ventral views.

Pimelodus telchitta Hamilton, 1822: 185, 378; type locality: Hooghly River at Kalna, West Bengal, India.

Glyptothorax telchitta: Darshan et al., 2015: 120 (valid).

Materials examined. RGUMF-183, 3 exs., 55.2–68.9 mm SL, Singen River, East Siang district, Arunachal Pradesh. RGUMF-unregistered, 50.1–65.3 mm SL, Dikrong River at Doimukh, Papum Pare district, Arunachal Pradesh.

Vernacular name in Arunachal Pradesh. *Ngurik* (in Nyishi), *Ngoling* (in Adi).

Diagnosis. Head and body with prominent tubercles, elliptical thoracic adhesive apparatus with narrow folds of skin and without a median depression, body surface lacks dark saddles on back, caudal peduncle depth with 4.7%–5.9% SL; and adipose-fin base length 9.5%–11.5% SL.

Coloration. In 10% formalin, dorsal and lateral surfaces of head and body brownish gray, fading to a lighter belly, no dark saddles on back. Distinct stripe running laterally along the dorsal second quarter of body, small black-brownish spots scattered throughout body surface. Dark brown patches appear in all fins but hyaline in the remaining portions of fins.

Distribution. Known from the Ganga-Brahmaputra drainage of India. In Arunachal Pradesh: Currently recorded from the Singen River in East Siang district.

Remarks. *Glyptothorax telchitta* and *G. botius* are very similar in appearance in having prominent tubercles on the head and body, a thoracic adhesive apparatus lacking a median depression, and a slender spindle shaped body.

Economic importance. This species consider as an ornamental fish.

Conservational status: The species is widely distributed throughout the Ganga-Brahmaputra basin only. IUCN status: Least concern (LC).

***Glyptothorax botius* (Hamilton 1822)**

Fig. 168.

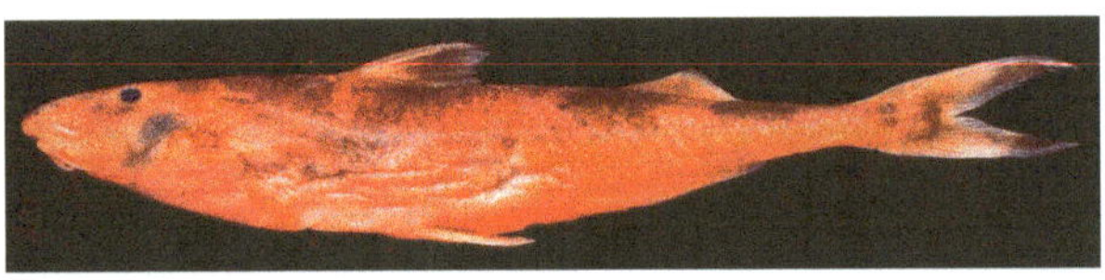

FIG. 168 *Glyptothorax botius*, RGUMF-186, 63.2 mm SL: lateral view.

Pimelodus botius Hamilton, 1822: 192, 378; type locality: Hooghly River at Kalna, West Bengal, India.

Glyptothorax botius: Darshan et al., 2015: 120 (valid).

Materials examined. RGUMF-440, 1 ex., 63.2 mm SL, Dikrong River at Doimukh, Papum Pare district, Arunachal Pradesh. RGUMF-186, 3 exs., 54.0–68.0 mm SL, Singen River in East Siang district, Arunachal Pradesh.

Vernacular name in Arunachal Pradesh. *Ngurik* (in Nyishi), *Ngoling* (in Adi).

Diagnosis. Head and body with prominent tubercles, thoracic adhesive apparatus without a median depression, narrow caudal peduncle with 3.1%–4.2% SL, presence of three dark brown saddles along dorsal surfaces of body, and adipose-fin base length with 12.0%–16.4% SL.

Coloration. In 10% formalin, dorsal surface of head body pale brown and lighter beneath. Dark brownish patches on occipital and opercular region. Three narrow dark brown saddles along the back: first on posterior part of dorsal fin, second on anterior quarter of adipose-fin base; and a third on caudal peduncle, immediately posterior to adipose-fin base. Ventral part of head and abdomen pale yellowish. Pectoral, pelvic, and anal fins hyaline. Caudal fin with a broad brown band along each lobe, remaining portion hyaline.

Distribution. Known from the Ganges and Brahmaputra river drainages in India and presumably also present in Bangladesh. In Arunachal Pradesh: see material examined.

Remarks. *Glyptothorax botius* readily distinguish from *G. telchitta* by its three broad saddles on back (vs. lacking saddles on back).

Economic importance. This species is considered an ornamental fish.

Conservational status: The species is widely distributed throughout the Ganga-Brahmaputra basin only. IUCN status: Least concern (LC).

***Glyptothorax indicus* Talwar 1991**

Fig. 169.

FIG. 169 *Glyptothorax indicus*, RGUMF-417, 125 mm SL: (A) lateral, (B) dorsal views, and (C) ventral view of thoracic adhesive apparatus.

Glyptothorax indicus Talwar, 1991: 654, Fig. 210; type locality: Streams of Terai, northern Bengal, India.

Glyptothorax indicus: Arunkumar, 2016: 182 (valid).

Material examined. RGUMF-417, 122–125 mm SL, Tezu River in Lohit district, Arunachal Pradesh.

Vernacular names in Arunachal Pradesh. *Ngurik* (in Nyishi), *Ngoling* (in Adi).

Diagnosis. Chevron-shaped thoracic adhesive apparatus with a narrowly pointed anterior apex at isthmus, presence of a patch of unculiferous striae behind the lower lip and between the bases of inner mandibular-barbel; distance between the posterior end of pectoral-fin base and the pelvic-fin origin is much greater than that of the distance between origins of pelvic and anal fins.

Coloration. In live, body with copper grayish with light brownish yellow on belly. Fins lightly copper tinged and caudal fin spotted with grayish. In 10% formalin, overall body yellowish brown superimposed with scattered dark brown irregularly shaped spots. Distal tip of dorsal, anal, and caudal fins hyaline.

Distribution. Known from the Ganga-Brahmaputra drainage of India. In Arunachal Pradesh: see material examined.

Economic importance. This species is considered an ornamental fish.

Conservational status: The species is widely distributed throughout the Ganga-Brahmaputra basin. IUCN status: Least concern (LC).

***Glyptothorax dikrongensis* Tamang & Chaudhry 2011**

Fig. 170.

FIG. 170 *Glyptothorax dikrongensis*, ZSI V/APFS/P-515, holotype, 94.2 mm SL: (A) lateral, and (B) ventral views.

Glyptothorax dikrongensis Tamang & Chaudhry, 2011: 2, Figs. 1, 3a–5a; type locality: Dikrong River at Doimukh, Arunachal Pradesh, northeastern India.

Glyptothorax dikrongensis: Arunkumar, 2016: 182 (valid).

Material examined. ZSI V/APFS/P-515, holotype, 94.2 mm SL; from Dikrong River at Doimukh, about 1 km upstream from Khola Camp RCC Bridge, Midpu, Arunachal Pradesh, northeastern India, June 28, 2009. ZSI V/APFS/P-516, 2 specimens, 78.0–85.3 mm SL; same data as holotype, July 8, 2009.

Vernacular names in Arunachal Pradesh. *Ngocher* (in Nyishi), *Ngoling* (in Adi).

Diagnosis. Chevron-shaped thoracic adhesive apparatus with a moderately broad anterior apex on the isthmus; presence of a patch of unculiferous striae on the behind the lower lip and between the bases of inner mandibular-barbel;

distance between the posterior end of pectoral-fin baseand the pelvic-fin origin is equal to the distance between the pelvic-fin origin and the anal-fin origin; presence of light to dark brown spots on the dorsal parts head and body.

Coloration. In live, pale yellowish body ground color, dorsal surface of head and body overlay with light to dark brown markings and spots. Dorsal-fin base dusky to light brown, adipose fin light brown, with hyaline subdistal margin. Pectoral-, pelvic-, and anal-fin bases dusky to light brown. Caudal-fin base with a vertical brownish band, its distal tip hyaline.

Distribution. Currently known only from Dikrong River at Doimukh; upstream from Khola Camp RCC Bridge, Midpu, Papum Pare district, Arunachal Pradesh.

Remarks. This species has very closed affinity with *Glyptothorax indicus*.

Economic importance. This species is considered as an ornamental fish.

Conservational status: The species is assumed to be restricted in the Dikrong river drainage only, a tributary of Brahmaputra basin. IUCN status: Not assessed (NA).

Glyptothorax striatus (McClelland 1842)

Fig. 171.

FIG. 171 *Glyptothorax striatus*, RGUMF-127, 197 mm SL: (A) lateral, and (B) ventral views.

Glyptosternon striatus McClelland, 1842: 587, pl. 6, Figs. 1 and 2; type locality: Kasyah [Kashya, Khasi] Hills, Meghalaya, India.

Glyptosternon striatus: Darshan et al., 2015: 120 (valid).

Materials examined. RGUMF-127, 02 exs., 73.0–197.0 mm SL, Tunaighat, Lohit district, Arunachal Pradesh, India. RGUMF-unregistered, 08 exs., 68.2–78.0 mm SL, Senkhi stream close to confluence point with Chimpu stream at Itanagar, Papum Pare district, Arunachal Pradesh, India.

Vernacular name in Arunachal Pradesh. *Ngurik* (in Nyishi), *Ngoling* (in Adi).

Diagnosis. Ventral surfaces of the pectoral spine and the first ray of pelvic fin with prominent plicae. Rhomboidal thoracic adhesive apparatus with wedge-shaped median depression, skin ridges absent in the depressed region; snout length 51.8%–54.7% HL, pectoral-fin length 18.7%–23.8% SL, long dorsal-fin spine with length 10.3%–15.7% SL, interdorsal length 24.9%–27.9% SL, body depth at anus 11.0%–14.7% SL, adipose-fin base length 10.7%–13.5% SL, caudal-peduncle length 18.4%–20.7% SL, caudal-peduncle depth 6.8%–8.6% SL.

Coloration. In 10% formalin, dorsal and lateral surfaces of head and body dark brown, light grayish brown on its ventral surface. A pale mid-dorsal stripe on the back, and another pale midlateral stripe on flank, superimpose above the lateral line. Midlateral stripe may be indistinct in larger specimens (see remark for detail). Adipose fin dark brown with light brown distal margin. Caudal fin brownish, scattered with dark brown spots and its distal tip hyaline.

Distribution. Known from both the Brahmaputra river drainage and the Surma-Meghna river system in India. In Arunachal Pradesh: see material examined.

Remarks. A pale midlateral stripe extending the entire length of body is present on the neotype (PUCMF 1058, 100.8 mm SL) of *G. Striatus*. Such pale midlateral stripe may be indistinct in larger specimen (e.g., RGUMF 127, 197 mm SL).

Economic importance. This species is considered as an ornamental fish.

Conservational status: The species is restricted in the Brahmaputra, Surma-Meghna river basin. IUCN status: Near threatened (NT).

***Glyptothorax pantherinus* Anganthoibi & Vishwanath 2013**
Fig. 172.

FIG. 172 *Glyptothorax pantherinus*, RGUMF-408, 120 mm SL: (A) lateral view, and (B) ventral view of thoracic adhesive apparatus.

Glyptothorax pantherinus Anganthoibi & Vishwanath, 2013: 2, Figs. 1 and 2; type locality: Noa Dehing River, Brahmaputra river basin, Deban-Namdapha, Changlang district, Arunachal Pradesh, India.

Glyptothorax pantherinus: Arunkumar, 2016: 182 (valid).

Material examined. RGUMF-408, 02 exs., 90.5–120 mm SL, Siang River at Pasighat, East Siang district, Arunachal Pradesh.

Vernacular name in Arunachal Pradesh. *Ngurik* (in Nyishi), *Ngoling* (in Adi).

Diagnosis. Ventral surface of the pectoral-fin spine and the first pelvic-fin ray plicated; leaf-shaped thoracic adhesive apparatus with no central depression; a bean-shaped yellowish spot one either side of the dorsal fin origin; body mottled with dark brown irregularly set ovoid blotches; predorsal length 31.3%–33.3% SL; caudal peduncle length 23.3%–25.7% SL; body depth at anus 15.2%–16.4% SL; and the first branchial arch bears 9 (2 + 7) gill rakers.

Coloration. In live, grayish yellow ground color mottled with dark gray ovoid blotches, several of which often conjoin to form irregular patches. Ventral surface of head and body in front of pelvic fin creamish to reddish yellow. Caudal fin mottled with dark gray ovoid blotches.

Distribution. Originally described from the Noa Dehing River, Changlang district of Arunachal Pradesh. The present study recorded this fish from the Siang River.

Economic importance. This species is considered as an ornamental fish.

Conservational status: The species is restricted in the upper Brahmaputra river basin. IUCN status: Not assessed (NA).

Glyptothorax mibangi **Darshan et al. 2015**

Fig. 173.

FIG. 173 *Glyptothorax mibangi*, RGUMF-243, holotype, 79.0mm SL: (A) lateral view, and (B) ventral of thoracic adhesive apparatus.

Glyptothorax mibangi Darshan et al., 2015: 116, Fig. 1; type locality: Arunachal Pradesh State, Tisa River near Longding, Brahmaputra River basin, India (valid).

Materials examined. RGUMF-243, holotype, 79.0mm SL; India: Arunachal Pradesh State, Tisa River near Longding, Brahmaputra river basin, 26°52′N, 95°21′E, D. N. Das & Party, October 18, 2008. RGUMF-244–245 (2), 75.9–83.5mm SL; collection data same as above.

Vernacular names in Arunachal Pradesh. *Ngurik* (in Nyishi), *Ngoling* (in Adi).

Diagnosis. Thoracic adhesive apparatus leaf-shaped with a spindle-shaped median depression, skin ridges present over the depressed region; ventral surface of pectoral spine and first pelvic-fin ray non-plaited; body slender with a depth 10.4%–13.5% SL; caudal peduncle shallow with depth 6.8%–8.3% SL; snout long with 52.9%–58.6% HL; and 9 (2 + 7) gill rakers on the first branchial arch. Dorsal fin with a spine and 6 branched rays, its spine serrated posteriorly with 4–6 serrae.

Coloration. In 10% formalin, dorsal and lateral surfaces of head and body uniformly brown, ventrally pale brown. A thin yellowish stripe runs mid-dorsally, another narrow pale yellow line superimposed on lateral line. Base and distal third-quarter of the dorsal and pectoral fins with dark brown bands. Adipose fin with a large dark brown blotch. Caudal-fin dark brown at base. Pelvic and anal fins with faint brown band on the midway between its base and distal tip.

Distribution. Known only from Tisa River at Longding, Longding district, Arunachal Pradesh.

Economic importance. This species consider as an ornamental fish.

Conservational status: The species is restricted in the upper Brahmaputra river basin. IUCN status: Not assessed (NA).

***Glyptothorax pasighatensis* Arunkumar 2016**

Fig. 174.

FIG. 174 *Glyptothorax pasighatensis*, RGUMF-407, 99.5 mm SL: lateral view.

Glyptothorax pasighatensis Arunkumar, 2016: 4, 2; type locality: Siang River at Pasighat, Arunachal Pradesh, India (valid).

Materials examined. RGUMF-407, 05 exs., 91.0–99.5 mm SL, Siang River at Pasighat, East Siang district, Arunachal Pradesh. RGUMF-unregistered, 5 exs., 80.8–85.7 mm SL, Chatjo River (R.) at Deomali, a tributary of Burhi-Dihing R., Brahmaputra basin (Bb.), Tirap district, Arunachal Pradesh (AP)

Vernacular names in Arunachal Pradesh. *Ngurik* (in Nyishi), *Ngoling* (in Adi).

Diagnosis. An oval-shaped thoracic adhesive apparatus with a median depression, ridges appears on the depressed region of the apparatus, adhesive apparatus width with 64.0%–70.4% its length. Mid-dorsal of body with 6–8 interdorsal and 3–5 post-adipose bumps formed over the distal tips of the underlying neural spines; caudal peduncle depth with 7.3%–9.3% SL; body depth at anus 13.5%–15.9% SL.

Coloration. Body clayish, a bean-shaped pale spot appears on either side of the dorsal-fin insertion origin; dorsal fin basally dark with a dark grayish subdistal dark band extending up to second branched ray. Caudal fin with a dark mark at the base and yellowish color tips.

Distribution. Known from the Siang River at Pasighat, East Siang district, Arunachal Pradesh.

Remarks. Based on the shaped and distribution of pleats on the thoracic adhesive apparatus, this species has very closed affinity with *G. mibangi* and *G. pantherinus.*

Economic importance. This species is considered an ornamental fish.

Conservational status: The species is restricted in the upper Brahmaputra river basin. IUCN status: Not assessed (NA).

72. Genus: *Oreoglanis* Smith, 1933

Oreoglanis Smith 1933: 70. Type species: *Oreoglanis siamensis* Smith 1933. Type by original designation (also monotypic).

Body elongate, depressed up to pelvic fin base; head strongly depressed, moderate in size, cover with soft skin; eyes small, subcutaneous; gill opening narrow, extending to middle of pectoral-fin base; mouth inferior, lip broad and flattened to form a sucker; lower lip with a continuous fold, often with a narrow notch in middle; upper jaw longer; teeth conical, no teeth on palate; barbels 4 pairs: 1 pair each of nasal, maxillary, outer and inner mandibular; dorsal fin with i,6 rays; adipose dorsal fin very long not confluent with caudal fin; no adhesive apparatus on thorax.

Key to species of *Oreoglanis*:

1a. Posterior margin of the maxillary barbel with a single row of villiform projections... *O. majuscula*

1b. Ventro-posterior margin of maxillary barbel with lobulate projections on the proximal third andlaciniate projections on the distal two-thirds ....*O. pangenensis*

Oreoglanis majuscula Linthoingambi & Vishwanath 2011
Fig. 175.

FIG. 175 *Oreoglanis majuscula*, RGUMF-128, 65.0mm SL: (A) dorsal, and (B) ventral views.

Oreoglanis majusculus Linthoingambi & Vishwanath, 2011: 61, Figs. 1a–c, 2 and 3, type locality: Kameng River at Rupa Brahmaputra basin, Arunachal Pradesh, India.

Oreoglanis majuscula: Chen et al., 2017: 107 (valid).

Material examined. RGUMF-128, paratypes, 7 exs., 35.0–75.0mm SL; Kameng River at Rupa, Arunachal Pradesh, India: 27°12′38″N; 92°24′06″E; Kenjum Bagra, August 21, 2005.

Vernacular name in Arunachal Pradesh. Ngoku (in Nyishi), *Ngoge* (in Adi), *Batpala* (in Monpa).

Diagnosis. Posterior margin of the maxillary barbel with a single row of villiform projections; lower lip modified into a labial fold with a shallow median notch; caudal-peduncle length of 18.0%–21.3% SL; pectoral-fin large and its distal margin reaching to anterior base of the pelvic fin; caudal fin with equal lobes.

Coloration. In 10% formalin, dorsal and lateral surfaces of head and body brownish, pale yellow on ventral surface of body. All fins brownish except a blackish blotch at the base of caudal fin.

Distribution. Known only from the Kameng River (Brahmaputra River drainage), in West Kameng district, Arunachal Pradesh, India.

Economic importance. This species is considered as an ornamental fish.

Conservational status: The species is restricted only in the upper reaches of Kameng River (Brahmaputra River Basin). IUCN status: Not assessed (NA).

***Oreoglanis pangenensis* Sinha & Tamang 2015**

Fig. 176.

FIG. 176 *Oreoglanis pangenensis*, ZSI/APRC P-791, holotype, 76.7mm SL: (A) lateral, (B) dorsal, and (C) ventral view of mouth part and belly.

Oreoglanis pangenensis Sinha & Tamang, 2015: 332, Figs. 1, 2b, 3 and 4; type locality: Pange River, upper Brahmaputra river basin, Lower Sunansiri district, Ziro Valley, Aro-Lenching, Apatani plateau, Arunachal Pradesh, India (valid).

Material examined. ZSI/APRC P-791, 76.7mm SL; Ziro Valley, Aro-Lenching, Pange River, a tributary of Brahmaputra River, Lower Subansiri district, 27°29.160′N 93°51.952′E, 1600m asl Arunachal Pradesh, India; B. Sinha & L. Tamang, June 9, 2013.

Vernacular names in Arunachal Pradesh. Ngoku (in Nyishi), *Ngoge* (in Adi), *Batpala* (in Monpa).

Diagnosis. Ventro-posterior margin of maxillary barbel with lobulate projections on the proximal third and laciniate projections on the distal two-thirds; lower lip margin entire with a distinct median notch; caudal fin emarginated; both the outer margin of caudal-fin lobe with a dark brown streak.

Coloration. In 10% formalin, body brownish to gray. Upper surface of head and body with cream patches. Dorsal-fin rays brownish and interradial membranes hyaline. Pectoral and pelvic fin light brown dorsally. Flank with faint whitish streak along above and below lateral line, lower streak more conspicuous than upper. Caudal-fin base dark brown, upper and lower lobe margins with a dark brown streak, distal margin brownish, and medially hyaline.

Distribution. Known only from Arunachal Pradesh: Pange River (headwater of Subansiri River), a tributary of the Brahmaputra river drainage, Lower Subansiri district.

Economic importance. This species is considered as an ornamental fish.

Conservational status: The species is restricted only in the Pange River (upper reaches of the Subansiri drainage) only, a tributary of Brahmaputra basin. IUCN status: Not assessed (NA).

73. Genus: *Parachiloglanis* Wu, He & Chu, 1981

Parachiloglanis Wu, He & Chu, 1981: 76, 79. Type species: Masc. *Glyptosternum hodgarti* Hora 1923. Type by original designation (also monotypic).

Body elongated, head depressed anteriorly; eyes small and subcutaneous; mouth inferior, bordered by papillated thick lips; no post-labial groove; teeth pointed in band on jaws, upper jaw uninterrupted; teeth absent on palate; barbels four pairs; gill opening restricted on sides of head and extended up to middle of base of pectoral fin; adipose dorsal fin long, low not confluent with caudal; anal fin very close to caudal fin but not reaching the base of caudal fin; caudal-fin margin lunate; no thoracic adhesive apparatus.

***Parachiloglanis hodgarti* (Hora 1923)**
Fig. 177.

FIG. 177 *Parachiloglanis hodgarti*, RGUMF-406, 75.0 mm SL: lateral view.

Glyptosternum hodgatri Hora, 1923: 38, pl. 2, Figs. 1–3, type locality: Pharping, Nepal.

Parachiloglanis hodgarti: Zhang et al., 2016: 176 (valid).

Material examined. RGUMF-406, 03 exs., 55.0–75.0 mm SL, Yomgo River at Mechuka, Upper Siang district, Arunachal Pradesh.

Vernacular name in Arunachal Pradesh. *Charba* (in Nyishi), *Ngobal* (in Adi), *Batpala* (in Monpa).

Diagnosis. No post-labial grove, lower lip continuous with the isthmus; lateral-line pores greatly reduced or absent; posterior margin of caudal fin indented to lunate, caudal fin with pale yellowish markings at the base and also to its distal tips; adipose fin low with depth 2.2%–4.3% SL, anal fin inserted very closed to caudal-fin base with preanal length 74.7%–81.9% SL.

Coloration. In live, dorsal and lateral surface of body brownish, dusky white on ventral surfaces up to vent. Caudal fin base with two whitish to yellowish blotches, one on upper half and the other below, both sometimes merged to form a whitish band at the base of caudal fin, distal tip of both lobes whitish.

Distribution. Known from India, Nepal, and Bangladesh. Reports of this species from China are likely to be a misidentification of an undescribed species. In Arunachal Pradesh: see material examined.

Remarks. The only congener of this species is *Parachiloglanis bhutanensis*, which was described from Bhutan recently.

Economic importance. This species is considered as an ornamental fish.

Conservational status: The species is widely distributed throughout the Ganga-Brahmaputra basin. IUCN status: Least concern (LC).

74. Genus: *Pseudecheneis* Blyth, 1860

Pseudecheneis Blyth 1860: 154. Type species: *Glyptosternon sulcatus* McClelland 1842. Type by monotypy.

Body subcylindrical; ventral surface flat up to pelvic fins; head short and depressed; eyes small, cover by a thin membrane; mouth inferior; lips thick and papillated; a broadly oval adhesive apparatus in chest, composed of series of transverse ridges.

Key to species of *Pseudecheneis*:

1a. The first dorsal-fin pterygiophore lacking a bony spur, interdorsal long 14.2%–21.4% SL .... *P. sulcata*

1b. The first dorsal-fin pterygiophore with a bony spur, interdorsal short 13.8%–14.1% SL..... *P. sirenica*

***Pseudecheneis sulcata* (McClelland 1842)**

Fig. 178.

FIG. 178 *Pseudecheneis sulcata*, RGUMF-132, 90.0 mm SL: dorsal view.

Glyptosternon sulcatus McClelland, 1842: 587, pl. 6, Figs. 1–3; type locality: Khasi Hills, Meghalaya, India.

Pseudecheneis sulcata: Gurung et al., 2013: 4884 (valid).

Material examined. RGUMF-132, 05 exs., 70–100 mm SL, Panye stream in Lower Subansiri district, Arunachal Pradesh.

Vernacular name in Arunachal Pradesh. *Kesari* (in Adi).

Diagnosis. The first dorsal-fin pterygiophore lacking a bony spur on the anterodorsal surface, thoracic adhesive apparatus with 12–14 transverse laminae; adipose-fin base length 17.8%–22.7% SL; interdorsal distance 14.2%–21.4% SL, pectoral fin with i,12–13 soft rays; anal iii,8–9 soft rays.

Coloration. In 10% formalin, body chestnut brownish on dorsal and lateral surfaces of head and body, fading to very light grayish on ventral surface. Dorsal surface of head and back with distinctive series of pale brown spots and bands. Dorsal and anal fins hyaline, with brownish basal and subdistal bands. Caudal fin brownish, with hyaline distal margin.

Distribution. This species is widely known from the Brahmaputra river drainage in India and Bangladesh. Reports from Tibet (China) need verification. In Arunachal Pradesh: see material examined.

Economic importance. This species is considered as an ornamental fish.

Conservational status: The species is widely distributed throughout the Ganga-Brahmaputra basin. IUCN status: Least concern (LC).

Pseudecheneis sirenica Vishwanath & Darshan 2007

Fig. 179.

FIG. 179 *Pseudecheneis sirenica*, MUMF-9070/1, holotype, 109.8 mm SL: (A) lateral, (B) ventral, and (C) first dorsal fin pterygiophore showing spur.

Pseudecheneis sirenica Vishwanath & Darshan, 2007: 2628, pl. 2, Fig. 2: type locality: Siren River, Upper Siang district, Brahmaputra drainage, Arunachal Pradesh, India.

Pseudecheneis sirenica: Anganthoibi & Vishwanath 2010: 199 (valid).

Materials examined. MUMF-9070/1, holotype, 109.8 mm SL, Siren River, Upper Siang district, Brahmaputra River drainage, Arunachal Pradesh, India. MUMF-9070/2–9070/3, paratypes, 5 exs., 87.0–102.2 mm SL, Siren River, Upper Siang District, Brahmaputra River drainage, Arunachal Pradesh, India.

RGUMF-131, 10 exs., 85.0–130.0mm SL, Siren River in Upper Siang District, Arunachal Pradesh.

Vernacular name in Arunachal Pradesh. *Ngurik* (in Nyishi), *Kesari* (in Adi).

Diagnosis. The first dorsal-fin pterygiophore with a prominent bony spur on the anterodorsal surface (Fig. 179C), thoracic adhesive apparatus with 14–16 transverse laminae; short interdorsal distance 13.8%–14.1% SL; adipose-fin base length 22.7%–23.1% SL; caudal peduncle thin with depth 4.1%–4.4% SL; first branchial arch with 12 (2 on upper arm + 10 on lower arm) gill rakers; and total vertebrae 36–37.

Coloration. In 10% formalin, dorsal and lateral surface of body dark brown and light brown beneath. Dorsal body surface with distinctive series of pale yellowish blotches: one ovate at dorsal origin, one circular on each side of the base of last dorsal-fin ray, one each on adipose origin and another at its posterior end; and the posteriormost one at the dorsal margin of caudal fin base. Adipose fin dark brown, with yellowish distal margin.

Dorsal and anal fin with yellowish band in the middle, distal margin hyaline. Caudal with two bands.

Distribution. This species was originally described from the Siren River (Brahmaputra drainage) in the Upper Siang district, Arunachal Pradesh. Recently, we encountered the species in the Yomgo River (upper reaches of the Siang River, Brahmaputra basin), and also in the Dibang River in Lower Dibang Valley District, Arunachal Pradesh.

Economic importance. This species is considered as an ornamental fish.

Conservational status: The species is restricted in the upper Brahmaputra basin only. IUCN status: vulnerable (VU).

18. Family: Erethistidae

Subfamily: Continae

75. Genus: *Conta* Hora, 1950

Conta Hora 1950: 194. Type species: Fem. *Pimelodus conta* Hamilton 1822. Type by original designation (also by monotypy and absolute tautonymy).

Body slightly elongated and subcylindrical; mouth small, snout bluntly pointed, eyes small subcutaneous; occipital spine narrowly not reaching to nuchal plate of dorsal fin; median longitudinal groove on dorsal surface of head reaching to base of occipital spine; humeral process support the edges abdomen; villiform teeth present on jaws; barbels four pairs; gill opening narrow, branchiostegal membrane confluent with isthmus; abdomen with plaited skin forming an elongated adhesive apparatus; dorsal and pectoral fin spine serrated on both edges; adipose fin small; caudal fin deeply forked with filamentous extension in upper lobe; body surface with numerous tubercules.

***Conta conta* (Hamilton 1822)**

Fig. 180.

FIG. 180 *Conta conta*, RGUMF-202. 49.0 mm SL: (A) lateral view, and (B) ventral part showing adhesive apparatus at belly.

Pimelodus conta Hamilton, 1822: 191, 378; type locality: Mahananda River, northeastern Bengal, India.

Conta conta: Shrestha, 2008: 168 (valid).

Material examined. RGUMF-202, 13 exs., 38.0–49.0 mm SL, Singen River near New Seren village, Arunachal Pradesh, India.

Vernacular name in Arunachal Pradesh. *Genikao* (in Galo).

Diagnosis. Elongated adhesive apparatus extending from gill opening to pelvic fin. Dorsal fin inserted almost opposite of the pelvic fin insertion, dorsal spine strong serrated in both the edges, 19–20 serrae on the anterior margin and 12–15 serrae on the posterior. Pectoral fin with strong spine and spine serrated on both edge, external serration with 28–29 serrae while internal serration with 17–18 serrae. Lateral line complete and overlay with prominent tubercles.

Coloration. In 10% formalin, body dark brownish-gray above, dirty white below. Fins with light bands, outer ray of caudal fin whitish border. Outer margin of dorsal, pectoral, pelvic, and anal fins with darker.

Distribution. Known from the Ganga-Brahmaputra river basin in India, Nepal, and Bangladesh. In Arunachal Pradesh: see material examined.

Remarks. *C. conta* differs from its nearest conger *C. pectinata* in having distally directed (vs. anteriorly directed or antrorse) serrations on the anterior edge of the pectoral spine.

Economic importance. This species is a popular ornamental fish.

Conservational status: The species is widely distributed throughout the Ganga-Brahmaputra basin. IUCN status: Data deficient (DD).

Subfamily: Erethistinae

76. Genus: *Erethistoides* Hora, 1950

Erethistoides Hora 1950: 190. Type species: *Erethistoides Montana* Hora 1950. Type by original designation.

Head and body strongly depressed; ventral surface flattened and straight; head pointed anteriorly; median longitudinal groove on head extending to the base of occipital spine; eyes small; mouth inferior; lips thin; barbels four pairs: 1 pair each of maxillary, nasal, inner, and outer mandibular; gill opening extending to ventral surface of head; nuchal plates of dorsal fin expanded on sides into wing-like extension; humeral process long and curved, narrowly separated from the pelvic bones; dorsal spine serrated anteriorly and pectinated posteriorly; pectoral spine with diverging serrations, antrorse on the distal half and retrorse on the proximal half; no thoracic adhesive apparatus.

***Erethistoides senkhiensis* Tamang et al. 2008**

Fig. 181.

FIG. 181 *Erethistoides senkhiensis*, ZSI/V/APRC/P-853, paratype, 38.9 mm SL: (A) lateral, and (B) ventral view of mouth part.

Erethistoides senkhiensis Tamang et al., 2008: 186; type locality: Senkhi stream, Arunachal Pradesh, India.

Erethistoides senkhiensis, Ng et al., 2012: 56 (valid).

Material examined. ZSI/V/APRC/P 853, 8, 36.7–38.9 mm SL, Senkhi stream close to confluence point with Chimpu stream, Itanagar, Arunachal Pradesh, India. Collected by L. Tamang, June 21 to August 15, 2007.

Vernacular name in Arunachal Pradesh. *Ngucher* (in Nyishi), *Nareng* (in Adi).

Diagnosis. Body with brownish background color, overlay with three broad blackish crossed bars: anterior one descends from the dorsal-fin base, middle band descents from the adipose-fin base and the last one on the caudal penducle. Median longitudinal groove on head long, extending from the level of anterior nares to the base of supraoccipital process; pectoral spine short with length 21.7%–27.9% SL; head length 24.3%–27.3%; body depth at anus 11.7%–16.4% SL; and posterior margin of caudal fin concave with longer lower lobe.

Coloration. See diagnosis section.

Distribution. Currently known from the Dikrong River and its tributaries (Brahmaputra basin) in Arunachal Pradesh, India.

Economic importance. This species is considered as an ornamental fish.

Conservational status: The species is restricted in the Dikrong River (Brahmaputra River Basin). IUCN status: Data deficient (DD).

77. Genus: *Erethistes* Müller & Troschel 1849

Erethistes (subgenus of *Bagrus*) Müller & Troschel 1849: 12. Type species: *Erethistes pusillus* Müller & Troschel 1849. Type by monotypy.

Body robust, covered with tubercles; abdomen flat; snout obtuse; nostril closed together, separated by a basal flap of nasal barbel; mouth inferior and narrow; lip thin, upper one papillated; eyes small; scapular on head and humeral process very prominent; gill opening, moderate and extend onto venter; barbels four pairs: 1 each of nasal, maxillary, outer and inner mandibulrs; dorsal spine serrated posteriorly, anteriorly smooth to granulated; pectoral spine serrated both on posterior and anterior margin, serration on posterior edge directed toward its base, and serrations on anterior edge either pointed toward its tip or in divergent pairs; no thoracic apparatus; paired fins non-plaited.

Key to the species of *Erethistes*:

1a. Serrae on the anterior edge of pectoral-fin spine divergent ....*E. pusillus*

1b. Serrae of the anterior edge of pectoral-fin spine not divergent 2

2a. Pectoral fin with eight branched rays, coracoid posterior process extending to the midpoint between the origins of pectoral and pelvic fins...........................*E. hara*

2b. Pectoral fin with five branched rays, coracoid posteriorprocess extending closed to pelvic-fin origin .... *E. jerdoni*

***Erethistes pusillus* Muller & Troschel 1849**
Fig. 182.

FIG. 182 *Erethistes pusillus*, RGUMF-134, 45.0mm SL: (A) dorsal, (B) lateral, and (C) ventral views.

Erethistes pussilus Muller & Troschel, 1849: 12, pl. 1, Fig. 2; type locality: Assam, India.

Erethistes pusillus: Thomson & Page 2006: 17 (valid).

Materials examined. RGUMF-134, 1 ex., 45.0 mm SL, Dikrong River at Doimukh, Papum Pare district, Arunachal Pradesh (AP). RGUMF-unregistered, 8 exs., 40.6–47.8 mm SL, Ranga River at Kimin and Senkhi River at Itanagar, Papum Pare district, AP.

Vernacular name in Arunachal Pradesh. *Ngucher* (in Nyishi), *Nareng* (in Adi).

Diagnosis. Dorsal fin with a strong spine and 5–6 branched rays, dorsal spine serrated posteriorly with 10–11 sarrae; pectoral fin with a strong spine and 5 branched rays, pectoral spine serrated in both edges, posteriorly with 9–11 serrae and anteriorly with 8–10 divergent serrae.

Coloration. In live, body ground color yellowish brown, superimposed with three broad dark brown bands.

Distribution. Known from India and Bangladesh. In Arunachal Pradesh: see material examined.

Economic importance. This species is considered as an ornamental fish.

Conservational status: The species is widely distributed throughout the Ganga-Brahmaputra basin. IUCN status: Least concern (LC).

***Erethistes hara* (Hamilton 1822)**

Fig. 183.

FIG. 183 *Erethistes hara*, RGUMF-136, 36.0 mm SL: lateral view.

Pimelodus hara Hamilton 1822: 190, 378; type locality: Hooghly River, south of Ranaghat, India.

Erethistes hara: Thomson & Page 2006: 15 (revision & valid).

Material examined. RGUMF-136, 1 ex., 36.0 mm SL, Dikrong River at Doimukh, Papum Pare district, Arunachal Pradesh.

Vernacular name in Arunachal Pradesh. *Ngucher* (in Nyishi); *Nareng* (in Adi).

Diagnosis. Pectoral fin with a spine and 8 branched rays, spine shorter than lateral head length and serrated on both edge, serrae of spine on anterior edge slanting toward its base; dorsal spine serrated along both edges and bears 5–6 soft rays, the posterior process of coracoid extending to the midpoint between the origins of pectoral and pelvic fins.

Coloration. In live, body chocolate brown with two light brown bars: one at mid-interdorsal and another at caudal peduncle, ventral surface of head and abdomen light brown, and distal margins of all rayed fins pale white. Caudal fin with an irregular W-shaped band subdistally.

Distribution. Known from the Ganga-Brahmaputra basin in India, Bangladesh, and Nepal. In Arunachal Pradesh: see material examined.

Economic importance. This species is considered as an ornamental fish.

Conservational status: The species is widely distributed in the Ganga-Brahmaputra and Barak-Meghna basins. IUCN status: Least concern (LC).

***Erethistes jerdoni* (Day 1870)**

Fig. 184.

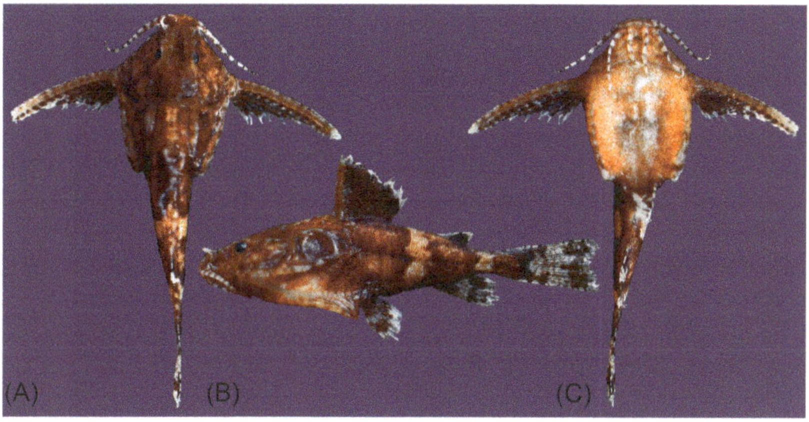

FIG. 184 *Erethistes jerdoni*, RGUMF-135, 27.0mm SL: (A) dorsal, (B) lateral, and (C) ventral views.

Hara jerdoni Day 1870: 39, pl. 4, Figs. 2a–c; type locality: Sylhet District, Bangladesh.

Erethistes jerdoni: Thompson & Page, 2006: 17 (valid).

Material examined. RGUMF-135, 1 exs., 27.0mm SL, Dikrong River at Doimukh, Papum Pare district, Arunachal Pradesh (AP).

Vernacular names in Arunachal Pradesh. *Ngucher* (in Nyishi), *Nareng* (in Adi).

Diagnosis. Pectoral spine length 42.2%–51.7% SL; total vertebrae 27–29; an emarginate caudal fin, long posterior process of coracoid reaching closed to pelvic-fin origin, pectoral fin branched rays 5 and occipital spine extending to the anterior tip of nuchal plate of dorsal fin.

Coloration. In live, body chocolate brown with two light brown bars: one at mid-interdorsal and another at caudal peduncle, ventral surface of head and abdomen light brown, and distal margins of all rayed fins and adipose fin with pale white.

Distribution. Known from the Ganga-Brahmaputra river basin in India and Bangladesh. Reported of this species from the Salween river drainage need confirmation. In Arunachal Pradesh: see material examined.

Remark. *Erethistes jerdoni*, *E. pusillus*, and *E. hara* are often encountered with other common fishes in Brahmaputra drainage in Assam and Arunachal Pradesh but discarded as weed fishes by the fishermen and fish sellers as the fish is not consumed as food.

Economic importance. This species is considered as an ornamental fish.

Conservational status: The species is widely distributed throughout the Ganga-Brahmaputra basin. IUCN status: Least concern (LC).

78. Genus: *Nangra* Day, 1877

Nangra Day 1877: 493. Type species: *Pimelodus nangra* Hamilton 1822. Type by absolute tautonymy.

Head depressed; snout margin slightly spatutate with short flat ventral portion and more or less markedly projecting anterior to jaws, mouth ventral and narrow; lips thick and fimbriated; villiform teeth on jaws, teeth absent on palate; barbels 4 pairs, 1 pair each of maxillary (having conspicuously developed posterior membrane), nasal, and 2 of mandibular; branchiostegal membranes free from the isthmus; dorsal fin with a spine and 6–10 branched rays, inserted above middle of pectoral fin; pectoral fin not reaching pelvic fin with 1 spine (serrated on both edges) and 8–10 rays; anal fin with 11–15 rays; caudal fin deeply forked and lateral line complete. Osteologically, the genus is diagnosed by its fan-shaped mesethmoid that is greatly expanded.

***Nangra assamensis* Sen & Biswas 1994**

Fig. 185.

FIG. 185 *Nangra assamensis*, RGUMF-176, 56 mm SL: lateral view.

Nangra assamensis Sen & Biswas, 1994: 441, Fig. 1; pl. 1; type locality: Brahmaputra River at Nimati Ghat, 14 km from Jorhat, Assam, India.

Nangra assamensis: Ferraris, 2007: 396 (valid).

Material examined. RGUMF-176; 5 exs., 54.0–56.0 mm SL, Dikrong River at Doimukh, Papum Pare district, Arunachal Pradesh.

Vernacular name in Arunachal Pradesh. *Nareng* (in Adi).

Diagnosis. Snout strongly projecting; median longitudinal groove on head extending to almost end of occipital process; nasal barbel short, not extending beyond the posterior margin of orbit; dorsal and pectoral-fin spines with filamentous extensions, dorsal fin with I,6 rays; pectoral spine serrated posteriorly with about 13–16 serrae.

Coloration. In live, body color muddy, dorsally darker; flank with a faint band along the lateral line.

Distribution. Known from the Ganga-Brahmaputra river basin in India and Nepal. In Arunachal Pradesh: see material examined.

Economic importance. This species is considered as an ornamental fish.

Conservational status: The species is widely distributed throughout the Ganga-Brahmaputra basin. IUCN status: Least concern (LC).

79. Genus: *Pseudolaguvia* Misra, 1976

Pseudolaguvia Misra 1976: 253. Type species: *Glyptothorax tuberculatus* Prashad & Mukerji 1929. Type by original designation (also monotypic).

Body moderately elongated; head slightly depressed, slender; snout obtuse; mouth inferior; lip thick, fleshy and papillated; nostril closed together, separated by a basal flap of nasal barbel; eyes small, thoracic adhesive apparatus with a median depression formed by the longitudinal skin folds; gill openings wide, gill membranes nearly meet one another on the venter; dorsal fin with a strong spine serrated posteriorly, anterior margin either smooth or granulate; anterior margin of pectoral spine serrated, all directing toward the tip of spine.

Key to the species of *Pseudolaguvia*:

1a. Mature body size large, reaching up to 47 mm SL; no bands on body.................. *P. magna*

1b. Mature body size lesser than 47 mm SL, bands on body....................................2

2a. Dorsal spine longer than 23.3% SL... *P. viriosa*

2b. Dorsal spine length shorter than 23.3% SL..3

3a. Adipose-fin base long with 15.9%–20.3% SL, dorsal-fin spine margins lack prominent serrae but rough on both the anterior and posterior edges, thoracic adhesive apparatus longer than its width; adipose fin short with length 15.4% SL; interorbital distance 23.4%–29.6% HL; pelvic fin reaching anal-fin origin...........................*P. shawi*

3b. Adipose-fin base long with 10.7%–14.2% SL, thoracic adhesive apparatus long and elliptical in shape, extending closed to the pelvic-fin origin; dorsal and pectoral-fin spine short with length 11.9%–15.0% SL and 16.6%–19.8% SL, respectively; total vertebrae 25–27....................................*P. jiyaensis*

Pseudolaguvia shawi **(Hora 1921)**

Fig. 186.

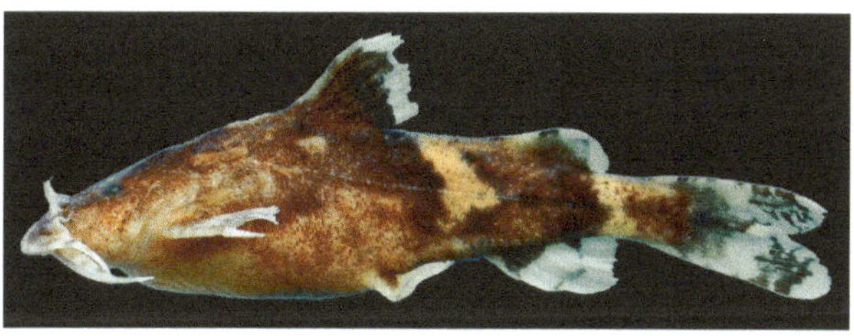

FIG. 186 *Pseudolaguvia shawi*, ZSI/V/APRC/-855, 28.6 mm SL: lateral view.

Laguvia shawi Hora, 1921: 740, pl. 29, Fig. 2; type locality: Mahanadi River below Darjeeling, West Bengal, India.

Pseudolaguvia shawi: Ng et al., 2016: 554 (valid).

Material examined. ZSI/V/APRC/-855, 28.6 mm SL, Senkhi River at Itanagar, Papum Pare district, Arunachal Pradesh.

Vernacular name in Arunachal Pradesh. *Ngucher* (in Nyishi), *Nareng* (in Adi).

Diagnosis. Dorsal-fin spine margins lacking prominent serrae but rough on both the anterior and posterior edges, thoracic adhesive apparatus longer than broad; adipose fin short with length 15.4% SL; interorbital distance 23.4%–29.6% HL; pelvic fin reaching anal-fin origin.

Coloration. In live, dorsal and lateral surfaces of body light to dark brown and whitish to yellowish on ventral surface; two broad vertical dark bands: one each below the base of dorsal and adipose fin present. Dorsal fin grayish with hyaline distal margin; pelvic fin hyaline with dark basal marking. All barbels with dark annulated markings.

Distribution. This species is known from the Ganga-Brahmaputra river drainage in India, Nepal, and Bangladesh. In Arunachal Pradesh: see material examined.

Economic importance. This species consider as an ornamental fish.

Conservational status: The species is widely distributed throughout the Ganga-Brahmaputra basin. IUCN status: Least concern (LC).

***Pseudolaguvia viriosa* Ng and Tamang 2012**

Fig. 187.

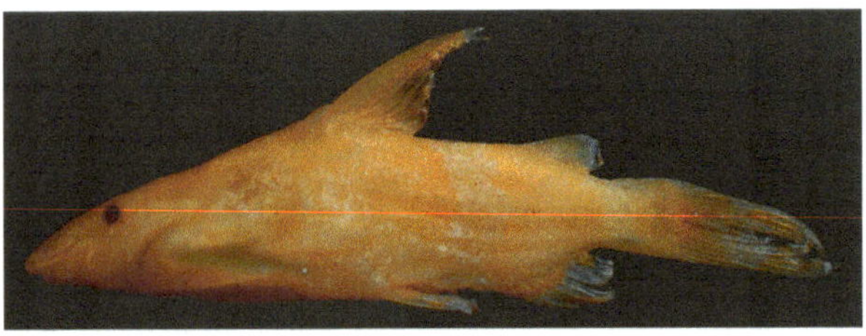

FIG. 187 *Pseudolaguvia viriosa*, ZSI V/APRC/P-524, paratype, 26.0 mm SL: lateral view.

Pseudolaguvia viriosa Ng & Tamang, 2012: 82, Fig. 1; type locality: Sille River from Ruksin and about 26 km before Pasighat, East Siang district, Arunachal Pradesh, India.

Pseudolaguvia viriosa: Tamang & Sinha, 2014: 38 (valid).

Material examined. ZSI V/APRC/P-524, 26.0 mm SL; India: Arunachal Pradesh, East Siang district, Sille River, approximately 1 km upstream from RCC bridge, about 10 km from Ruksin and about 26 km before Pasighat, 27°52′37.6″N 95°18′18.0″E; L. Tamang, October 28–November 2, 2011.

Vernacular names in Arunachal Pradesh. *Ngucher* (in Nyishi), *Nareng* (in Adi).

Diagnosis. Thoracic adhesive apparatus elliptical in shape with a median depression, and extending to midway between the base of last pectoral-fin ray and the pelvic-fin origin; very long dorsal and pectoral spine with length 23.4%–29.0% SL and 26.9%–32.9% SL, respectively, adipose-fin base length 12.5%–15.4% SL; body depth at anus 16.9%–19.0% SL; caudal peduncle length 14.8%–17.7% SL.

Coloration. In 70% alcohol, dorsal and lateral sides of head and body brownish, beneath creamish. Pale spots overlay on tip of occipital spine, lateral tips of nuchal plates and posterolateral tip of Weberian lamina; creamish band encircling the body by passing between the dorsal fin and the adipose fin dorsally and between the anal fin and the pelvic fin below. Caudal peduncle with a creamish blotch or band on sides, elongated creamish patched on dorsal and ventral margins. Dorsal fin brownish, distal margin hyaline; pelvic and anal fin banded. Caudal fin with a large brown band at the distal third, distal tips of the upper and the lower lobe hyaline.

Distribution. Known only from the Sille River (a tributary of the Siang River) in Arunachal Pradesh, India.

Economic importance. This species is considered as an ornamental fish.

Conservational status: The species is restricted in the Sille River, a tributary of the Siang River (Brahmaputra river basin). IUCN status: Not assessed (NA).

***Pseudolaguvia jiyaensis* Tamang & Sinha 2014**

Fig. 188.

FIG. 188 *Pseudolaguvia jiyaensis*, ZSI/V/APRC/P-1034, holotype, 29.9 mm SL: (A) lateral, and (B) dorsal views.

Pseudolaguvia jiyaensis Tamang & Sinha, 2014: 45, Figs. 5 and 6; type locality: Arunachal Pradesh, Lower Dibang Valley District, Jiya stream, near Bolik village from Roing toward Shantipur, Assam, India (valid).

Material examined. ZSI/V/APRC/P-1034, holotype, 29.9 mm SL; Jiya stream, near Bolik village, Lower Dibang Valley District Arunachal Pradesh, India, approximately 14 km from Roing toward Shantipur, Assam, 28°00.377′N 95°45.562′E; 149 m asl., B. Sinha & L. Tamang, March 18, 2014.

Vernacular names in Arunachal Pradesh. *Ngucher* (in Nyishi), *Nareng* (in Adi).

Diagnosis. Body brownish with creamy vertical bands, one between dorsal and adipose fins and the other on the caudal peduncle, the anterior band twice as broad as the posterior. Thoracic adhesive apparatus long and elliptical in shaped, extending closed to the pelvic-fin origin; dorsal and pectoral-fin spine short with length 11.9%–15.0% SL and 16.6%–19.8% SL, respectively; total vertebrae 25–27.

Coloration. In 10% formalin, body brownish, overlay with two creamy vertical bands. Dorsal fin brownish with distal margin hyaline. Pelvic fin with subdistal streak and sometimes absent. Caudal fin with grayish base extending up to median rays. Pectoral, pelvic, and anal fins with brown bases and transverse subdistal bands.

Distribution. Known only from the Jiya stream in Lower Dibang Valley district, Arunachal Pradesh, India.

Remarks. *P. jiyaensis* distinguished readily from *P. magna* by its shorter interdorsal distance.

Economic importance. This species is considered as an ornamental fish.

Conservational status: The species is restricted in the Jiya stream, a tributary of the Dibang River (Brahmaputra river basin). IUCN status: Not assessed (NA).

***Pseudolaguvia magna* Tamang & Sinha 2014**
Fig. 189.

FIG. 189 *Pseudolaguvia magna*, ZSI/V/APRC/P-947, holotype, 46.1 mm SL: lateral view.

Pseudolaguvia magna Tamang & Sinha, 2014: 38, Figs. 1 and 4; type locality: Arunachal Pradesh, Lower Dibang Valley district, Jiya stream, near Bolik village, from Roing toward Shantipur, Assam, India.

Material examined. ZSI/V/APRC/P-947, 1 ex., 46.1 mm SL; Lower Dibang Valley District, Jiya stream, near Bolik village Arunachal Pradesh, India, approximately 14 km from Roing toward Shantipur, Assam, 28°00.377′N 95°45.562′E; 149 m asl., B. Sinha & L. Tamang, March 18, 2014.

Vernacular names in Arunachal Pradesh. *Ngucher* (in Nyishi), *Nareng* (in Adi).

Diagnosis. Large mature body size, reaching up to 47 mm SL; no bands on lateral surface of body; rhomboidal thoracic adhesive apparatus, slightly longer than broad, extending from the isthmus to the level of tip of last pectoral-fin ray; pelvic-fin origin almost perpendicular to the base of last dorsal fin ray. Pectoral fin with a strong spine, and 6 soft rays; pelvic fin with i,5 rays and anal fin with iii–iv,4–5 rays.

Coloration. In 10% formalin, dorsal surface of head brownish, darker on occipital region. Body brownish above, fading to light brown on the sides, bars absent on the flank. Mid-dorsal stripes appear along the interdorsal space and also between the adipose and caudal fins. Dorsal fin brownish with a hyaline central streak, distal margin hyaline. Pelvic and pectoral fins dusky basally with dark brown subdistal band, distal margin hyaline. Caudal-fin dark brown, superimpose with hyaline spots, distal tip both lobes hyaline.

Distribution. Known only from Jiya stream in Lower Dibang Valley district, Arunachal Pradesh, India.

Remarks. This is the largest species among its congeners in Brahmaputra drainage.

Economic importance. This species is considered as an ornamental fish.

Conservational status: The species is restricted in the Jiya stream, a tributary of the Dibang River (Brahmaputra River basin). IUCN status: Not assessed (NA).

19. Family: Clariidae

80. Genus: *Clarias* Scopoli, 1763

Clarias Scopoli (ex Gronow) 1777: 455. Type species: *Silurus anguillaris* Linnaeus 1758. Type by subsequent designation.

Body elongated and compressed; head greatly depressed, covered with osseous plates dorsally; mouth terminal and wide; lip fleshy and papillated; jaws subequal, upper longer; barbels 4 pairs; gill membrane deeply notched, partly joined but free from isthmus; a suprabranchail accessory respiratory organ present; skin smooth; dorsal fin without spine, long with 62–77 rays; pectoral fin with a strong spine with 7–11 branched rays; anal fin long with 45–63 rays; caudal fin almost rounded and not confluent with dorsal or anal fin.

Key to the species of *Clarias*:

1a. Pectoral spine serrated on both edges.. *C. magur*

1b. Pectoral spine serrated only on anterior margin .... *C. gariepinus*

***Clarias magur* (Hamilton 1822)**
Fig. 190.

FIG. 190 *Clarias magur*, RGUMF-398, 170 mm SL: lateral view.

Macropteronotus magur Hamilton, 1822: 146, 374; type locality: Ganges River, India.

Clarias magur: Kottelat, 2013: 243 (valid).

Material examined. RGUMF-398, 01. ex., 170 mm SL, Purchased from Doimukh bazar, Papum Pare district, Arunachal Pradesh.

Vernacular name in Arunachal Pradesh. *Magur* (in Nyishi & Galo), *Beyek* (in Adi).

Diagnosis. Head large and depressed, snout wide with rounded lateral margin. Dorsal and fin commencing from near occiput and extending to but not continuous with caudal fin. Pectoral fin with 1 spine and 8–9 soft rays. Pectoral spine serrated on both edges, strongly serrated on the entire length of anterior margin with 26–29 serrae, and posterior margin weakly serrated on its distal half. Length of occipital spine nearly half of its wide at base. Anal fin long, and not confluent with caudal fin.

Coloration. Rich reddish brown or grayish black, smooth skinned. In live, body uniformly grayish to dark gray above, fading to pale gray beneath. Dorsal, pectoral, anal and caudal fins gray to dark-gray. Pelvic fin hyaline.

Distribution. Native in the Ganga-Brahmaputra basin in India, Bangladesh, and Nepal. In Arunachal Pradesh: occurring in all major rivers of the state.

Remarks. Earlier records of *Clarius batrachus* form the Ganga-Brahmaputra basin in Northeast India are misidentifications of *C. magur*.

Economic importance. This species is a highly demanded popular food fish.

Conservational status: The species is widely distributed throughout the Ganga-Brahmaputra basin. However, its population has been drastically reduced in the last decade due to overexploitation. Induced breeding of this species has not been employed successfully in most of the government and private fish breeding farms due to complicated protocol so far developed by various research institutions. IUCN status: Endangered (EN).

***Clarias gariepinus* (Burchell 1822)**
Fig. 191.

FIG. 191 *Clarias gariepinus*, RGUMF-unregistered, 293 mm SL: dorsolateral view.

Silurus (Heterobranchus) gariepinus Burchell 1822: 425, Fig. p. 445; type locality: Vaal River, Smidtsdrift, above confluence with Riet River, Cape Province, South Africa.

Clarias gariepinus: Koerber et al. 2017: 4 (valid).

Material examined. RGUMF-unregistered, 01 ex., 293 mm SL, purchased from Naharlagun Bazar, Papum Pare district, Arunachal Pradesh.

Vernacular names in Arunachal Pradesh. *Magur* (in Nyishi), *Beyek* (in Adi).

Diagnosis. Pectoral spine robust, anterior margin serrated, posteriorly smooth. Length of occipital spine almost equal to its wide at the base. Long dorsal and anal fins with 61–80 and 45–65 soft rays, respectively, spine absent on both fins. Dorsal and anal fins not continuous with caudal fin.

Coloration. Generally, two color patterns known for this species: a uniform pattern and a variegated pattern. In uniform pattern: dorsal and lateral surfaces of head and body dark grayish to greenish black; belly and abdomen lightly colored. In variegated pattern: live specimens show irregular ash color blotches on light grayish body background. Belly and ventral parts of the paired fins are whitish.

Distribution. Native in Africa; introduced elsewhere including India. In Arunachal Pradesh: illegally culture in ponds.

Remarks. A highly carnivorous catfish.

Economic importance. This species is a food fish.

Conservational status: The species is widely distributed and also introduced elsewhere legally or illegally. IUCN status: Least concern (LC).

20. Family: Heteropneustidae

81. Genus: *Heteropneustes* Müller, 1840

Heteropneustes Müller 1840: 115. Type species: *Silurus fossilis* Bloch 1794. Type by monotypy.

Body elongated and compressed; head strongly depressed; snout flat, mouth small and terminal, upper jaw slightly longer; lip fleshy and papillated; dorsal fin short with 6–8 rays, lacking spine; pectoral fin with a strong serrated (inner) spine with venom glands; lateral line complete; barbel 4 pairs; adipose fin absent; anal fin long and reaching the caudal fin; gill opening wide, separated by a deep notch free from isthmus; gill cavity with an accessory two tubular air sacs extending posteriorly into the caudal peduncle, caudal fin rounded and not confluent from anal fin.

***Heteropneustes fossilis* (Bloch 1794)**

Fig. 192.

FIG. 192 *Heteropneustes fossilis*, RGUMF-unregistered, 73.8 mm SL: lateral view.

Silurus fossilis Bloch, 1794: 46, pl. 370, Fig. 2; type locality: Tranquebar (Tharangambadi) India.

Heteropneustes fossilis: Kottelat, 2013: 245 (valid).

Materials examined. RGUMF-140, 12 exs., 100–210 mm SL, Dikrong River at Doimukh, Papum Pare district, Arunachal Pradesh (AP). RGUMF-unregistered specimens: 15 exs., 112.5–189.6 mm SL, Siang River in East Siang district; Subansiri River in Lower Subansiri district and Ranga River in Papum Pare district, AP.

Vernacular name in Arunachal Pradesh. *Singhi* (in Nyishi), *Beyek* (in Adi).

Diagnosis. Body elongated and slightly compressed. Skin smooth. Dorsal fin very short with 6–7 rays. Adipose dorsal fin absent. Pectoral with 1 spine and 7 rays, spine strong, serrated posteriorly and also with a few outward serrations anteriorly. Anal fin long with 64–70 rays; caudal fin round with 18–19 rays and separated from the anal fin.

Coloration. In live, body reddish brown or purplish brown but in mature stage of specimens it is blackish.

Distribution. Known from India, Bangladesh, Lao People's Democratic Republic, Myanmar, Nepal, Pakistan, Sri Lanka, and Thailand. In Arunachal Pradesh: see material examined.

Remarks. Commonly known as "stinging catfish" because of its ability to deliver a painful sting to humans.

Economic importance. This species is a highly demand popular food fish.

Conservational status: The species is widely distributed in wetland water bodies throughout the Ganga-Brahmaputra basin. IUCN status: Least concern (LC).

VI. ORDER: SALMONIFORMES

Salmonidae is the only family under this order. The order is characterized by its deep posterior myodome with eye musculature passing through and attaching to trunk muscles; presence of adipose fin, mesocoracoid and pelvic axillary process; gill membranes free from isthmus and extending far forward; vertebral centra pitted and the last three vertebrae turned up; vertebrae 50–75; 11–210 pyloric caeca; branchiostegal rays 7–20. Maximum length known up to 1.5 m.

21. Family: Salmonidae

Subfamily: Salmoninae

82. Genus: *Oncorhynchus* Suckley, 1861

Oncorhynchus (subgenus of *Salmo*) Suckley 1861: 313. Type species: *Salmo scouleri* Richardson 1836. Type by original designation.

Premaxillaries considerably elongated in adult males, its tip decurved, extending conspicuously beyond the extremity of lower jaw. Upon the tip of lower jaw there is a knob, more or less broad and heavily armed with strong curve teeth, as of premaxillaries above (Suckley, 1861).

***Oncorhynchus mykiss* (Walbaum 1792)**
Fig. 193.

FIG. 193 *Oncorhynchus mykiss*, RGUMF-443, 98.4 mm SL: lateral view.

Salmo mykiss Walbaum, 1792: 59; type locality: Kamchatka, Russia.
Oncorhynchus mykiss: Zhang et al., 2016: 212 (valid).

Materials examined. RGUMF-166, 5 exs., 58–108 mm SL, Poma River at Ramgath, Papum Pare district, Arunachal Pradesh (AP). RGUMF-443, 01 ex., 98.4 mm SL, Sela lake at Selapass, Tawang district, AP.

Vernacular names in Arunachal Pradesh. Not available (in Nyishi, Adi & Galo).

Diagnosis. Body elongated, spindle shaped and somewhat compressed especially in larger fish. Dorsal fin with 3–4 spines and 10–12 soft rays; anal fin with 3–4 spines and 8–12 soft rays; caudal fin with 19 rays. Midlateral row with 115–130 scales. Extensive pink to red dotted from head to caudal base and adipose fin present, usually with black edge.

Coloration. Live body coloration varies with habitat, size, and sex. Body surface bluish to greenish above, pinkish to reddish spots along the lateral line and silver below. Head, flanks, adipose, and dorsal fin spotted.

Distribution. Known from India, Bangladesh, Sri Lanka, Pakistan, and Myanmar. In Arunachal Pradesh: introduced in Tawang and West Kameng district.

Remarks. A highly carnivorous introduced fish.

Economic importance. A highly prized food fish; also regarded as game fish.

Conservational status: The species is widely distributed throughout the multiple basins of Indian Ocean. IUCN status: Not assessed (NA).

83. Genus: *Salmo* Linnaeus, 1758

Salmo Linnaeus 1758: 308. Type species: *Salmo salar* Linnaeus 1758. Type by subsequent designation.

Body somewhat lengthened, mouth large, and supplied with teeth, which are conical, pointed, and formidable, implanted into all bones of the mouth, into the interior maxillary bones, both superior and inferior, the vomer and plates bones, into the tongue, and into the branchial arches.

***Salmo trutta* Linnaeus 1758**

Fig. 194.

FIG. 194 *Salmo trutta*, RGUMF-392, 170 mm SL: lateral view.

Salmo trutta, Linnaeus 1758: 308, type locality: European rivers.

Salmo trutta: Kottelat & Freyhof 2007: 408 (valid).

Material examined. RGUMF-392, 170, 01 ex., 125.5 mm SL, Tawang Trout hatchery, Govt. of Arunachal Pradesh.

Vernacular names in Arunachal Pradesh. Not available (in Nyishi, Adi, & Galo).

Diagnosis. Rayed dorsal fin inserted above pelvic fins with 12–14 rays. Adipose dorsal smooth, small, inserted above mid-anal. Anal fin with 8–12 rays. First gill arch with 17–20 rakers on the lower arm. Lateral line scales 115–130. Head, body, dorsal fin, adipose, and caudal fins with numerous scattered blackish spots.

Coloration. In live, body background color grayish yellow, superimpose with numerous scattered blackish spots. Dorsal, adipose and caudal fins spotted.

Distribution. Introduced widely elsewhere; native in Eastern-north Atlantic, Baltic Sea, North Sea, Mediterranean Sea, Black Sea, Caspian Sea, Europe, and western Asia. In Arunachal Pradesh: introduced in Tawang district and West Kameng district.

Remarks. A highly carnivorous fish.

Economic importance. A highly prized food fish; also regarded as a game fish.

Conservational status: The species is widely distributed and also introduced elsewhere. IUCN status: Least concern (LC).

VII. ORDER: GOBIIFORMES

This order is formed by two subgroups: earlier Kurtiformes and formerly known suborder Gobioidei in the order Perciformes. The order Gobiiformes is characterized by the absence of parietals, lachrymal ossified and other element in the infraorbitals series are unossified or absent, lateral line system reduced to variably developed cephalic canals (except in primitive Rhyacicthys); swim bladder usually absent; gill membranes usually attached to isthmus; barbels may present in some forms; no pyloric caeca; 4–10 flexible spines in the spinous first dorsal; first element of second dorsal and anal fins usually spinous with variable number of posterior soft rays; pelvic fins beneath pectoral fin, with a spine and 4–5 soft rays, usually joined in the middle; unique sperm gland; 25–35 vertebrate.

22. Family: Gobiidae

84. Genus: *Glossogobius* Gill, 1859

Glossogobius Gill 1859: 46. Type species: *Gobius platycephalus* Richardson 1846. Type by monotypy.

Body elongated, anteriorly cylindrical and compressed posteriorly; head depressed; no scales on check and operculum; mouth oblique, not extending beneath anterior rim of orbit; eyes large, superior; lips thick, jaws with villiform teeth in several rows; tongue bilobate; gill openings extending far forward below in advance of orbit; pelvic fins oblong joined mid-ventrally; two dorsal fins, separated by distinct interspace; first dorsal with 6 rays, its origin located above distal half of pectoral fin, and the second with 6–10 rays; pectoral-fin base scaled; pelvic fins joined in the middle; anal fin with 8–9 rays; caudal fin oblong or round; scales ctenoid on body; cycloid on head

***Glossogobius giuris* (Hamilton 1822)**
Fig. 195.

FIG. 195 *Glossogobius giuris*, RGUMF-154, 103 mm SL: lateral view.

Gobius giuris Hamilton, 1822: 51, 366; type locality: all the ponds and fresh water rivers of the Gangetic provinces, India.

Glossogobius giuris: Kottelat, 2013: 407 (valid).

Materials examined. RGUMF-154, 10 exs., 103–120 mm SL, Ranga River at Kimin, Papum Pare district, Arunachal Pradesh (AP). RGUMF-unregistered specimens: 8 exs., 116–126.8 mm SL, Ranga and Dikrong River at Doimukh, Papum Pare district, AP.

Vernacular names in Arunachal Pradesh. Not available (in Nyishi, Adi, & Galo).

Diagnosis. Snout pointed; lower jaw prominent and slightly longer than the upper jaw. Pelvic fin joined. Body with 4–6 dark blotches along the lateral line. The lateral line runs straight in the middle of the flank. Dorsal fins separated: first dorsal fin with vi rays, second with i,9 rays. Pectoral fin with ii,19–21 rays. Pelvic fin with ii,7 rays. Anal fin with i,8 rays and caudal fin rounded with 12–14 rays.

Coloration. In live, body yellowish brown with some shine of gold on the sides. On the dorsal and flanks there are numerous blackish spots, forming large irregular spots. The second dorsal fin, and caudal fin are spotted. Pectoral, pelvic and caudal are light grayish or often hyaline.

Distribution. Widely distributed, India: throughout the country in all major river basins, Bangladesh, Pakistan, Sri Lanka, East Africa, Malay Peninsula, China, Japan, Philippines, and South Pacific islands. In Arunachal Pradesh: occurring in all major rivers.

Remarks. This fish can migrate through sea.

Economic importance. A food fish.

Conservational status: The species is widely distributed (see distribution section). IUCN status: Least concern (LC).

23. Family: Ambassidae

85. Genus: *Chanda* Hamilton, 1822

Chanda Hamilton 1822: 103, 370. Type species: *Chanda nama* Hamilton 1822. Type by designation of ICZN.

Body short, deep, and strongly compressed; semitransparent body; head short and compressed; abdomen rounded; snout pointed, mouth wide and oblique, extending beneath the level of anterior rim of the eyes or slightly beyond; eyes large; lips thin, lower jaw longer than upper; villiform teeth on jaws, palate, and tongue; lower limb of preopercle with double serrated edge; 2 dorsal fins, continuous, the first with 7 spines, and the second with 1 simple and 9–18 rays; a horizontal anteriorly directed spine present in front of the dorsal fin; anal fin with 3 spines followed by 9–17 rays; lateral line present; caudal fin forked; scales cycloid, small to moderate in size.

***Chanda nama* Hamilton 1822**

Fig. 196.

FIG. 196 *Chanda nama*, RGUMF-146, 77 mm SL: lateral view.

Chanda nama Hamilton, 1822: 109, 371, pl. 39, Fig. 37; type locality: ponds throughout Bengal, India.

Chanda nama: Kottelat, 2013: 321 (valid).

Material examined. RGUMF-146, 14 exs., 45.0–77.0 mm SL, Ranga River at Kimin, Papum Pare district, Papum Pare district, Arunachal Pradesh.

Vernacular names in Arunachal Pradesh. Not available (in Nyishi, Adi, & Galo).

Diagnosis. Body longitudinally ovate, nearly transparent and strongly compressed. Mouth large with prominent lower jaw. Scales minute with 112–116 lateral line scales. First dorsal fin with 7 spines and the second with i,14–16 rays, anal fin with 3 spines and 17 rays.

Coloration. In live, body silvery with sparsely scattered minute blackish dots on body. Caudal fin dusky yellowish with pale distal margin.

Distribution. Known from India, Nepal, Pakistan, and Bangladesh. Records from Myanmar and Malaysia are probably erroneous. In Arunachal Pradesh: Dikrong, Subansiri, Senkhi, and Ranga rivers.

Remarks. Only this species is known under the genus *Chanda*.

Economic importance. This species is a popular ornamental fish due to its small size and distinctive coloration.

Conservational status: This fish is wide spread species occurring in the Ganga-Brahmaputra river basin. IUCN status: Least concern (LC).

86. Genus: *Parambassis* Bleeker, 1874

Parambassis Bleeker 1874: 86, 102. Type species: *Ambassis apogonoides* Bleeker 1851. Type by original designation.

Body elongated and strongly compressed; translucent to transparent body; mouth large and gape oblique; extending beneath the anterior rim of orbit; supraorbital ridge with one or two spines posteriorly; preorbital bone serrated on both ridge and edge; tongue toothless; two dorsal fins which are continuous, the first with 7–8 spines, and the second with 1 spine followed by about 9–16 rays; lateral line series with 30–100 scales; 9–17 anal-fin rays, 24–25 vertebrae.

Key to the species of *Parambassis*:

1a. Two black spots on either side of the anal-fin origin which are connected by a blackish curved line..................... *P. bistigmata*
1b. No spots on anal-fin origin...2
2a. Lower arm of first branchial arch with 12–18 rakers; lateral line complete with 80–100 scales .... *P. baculis*
2b. Lower arm of first branchial arch with more than 20 rakers; lateral line complete with 55–60 scales ....*P. ranga*

Parambassis ranga **(Hamilton 1822)**
Fig. 197.

FIG. 197 *Parambassis ranga*, RGUMF-147, 40.0 mm SL: lateral view.

Chanda ranga Hamilton, 1822: 113, 371, pl. 6. Fig. 38; type locality: Gangetic provinces, India.

Parambassis ranga: Kottelat, 2013: 322 (valid).

Materials examined. RGUMF-147, 10 exs., 30–73 mm SL, Ranga River at Kimin, Papum Pare district, Arunachal Pradesh (AP). RGUMF-unregistered, 04 exs., 42–70.6 mm SL, Subansiri River in Lower Subansiri district, AP. RGUMF-unregistered, 06 exs., 36–68.2 mm SL, Ranga River in Lower Subansiri district, AP.

Vernacular names in Arunachal Pradesh. Not available (in Nyishi, Adi, & Galo).

Diagnosis. Body short, deep, and strongly compressed. Mouth small and oblique, minute teeth on both jaws; lateral line complete with 55–60 scales. First and second dorsal fin joined, first dorsal with 7 spines and second dorsal with a spine and 13–14 rays, pectoral fin with 12–13 rays, anal fin with 3 spines and 13–15 rays.

Coloration. In live, this species has a striking transparent body revealing its bones and internal organs; the male develops a dark edge to the dorsal fin.

Distribution. Known from India, Bangladesh, Nepal, and Pakistan, Myanmar, Malaysia (Peninsular Malaysia), and Cambodia. In Arunachal Pradesh: see material examined.

Remarks. This fish is commonly known as the Indian glassy fish. It has high demand in the aquarium fish trade.

Economic importance. This species is a very popular ornamental fish.

Conservational status: A wide spread species occurring in the multiple basin of Southeast Asia. IUCN status: Least concern (LC).

***Parambassis bistigmata* Geetakumari 2012**

Fig. 198.

FIG. 198 *Parambassis bistigmata*, RGUMF-422, 44.1 mm SL: (A) lateral, and (B) ventral views.

Parambassis bistigmata Geetakumari 2012: 60, Figs. 1 and 3; type locality: Ranga River, Kimin Station, Brahmaputra drainage, Arunachal Pradesh, India.

Parambassis bistigmata: Mayanglambam & Vishwanath, 2015: 587 (valid).

Material examined. RGUMF-422, 1 ex., 44.1 mm SL, Ramasikri River, a tributary of Siang River, East Siang district, Arunachal Pradesh.

Vernacular names in Arunachal Pradesh. Not available (in Nyishi, Adi, & Galo).

Diagnosis. Body small and compressed; 8 spines on first dorsal and 11–12 soft rays on second dorsal fins; 3 spines and 11–12 soft rays on anal fin; 1 spine and 5 soft rays on pelvic fin; and 10 soft pectoral fin rays. Two black spots on either side of the anal-fin origin (Fig. 198B), joined by a fine black curved line.

Coloration. In 10% formalin, body ground color brownish with a faded blackish midlateral line on posterior half of body. Two black spots on either side of the anal-fin origin, joined the two by a fine black curved line. The interradial membrane between the second and fourth spines of first dorsal with a distinct darkish outer margin, second dorsal hyaline. Pectoral, pelvic and anal fins hyaline.

Distribution. Known from in Arunachal Pradesh (Ranga and Subansiri rivers) and also from the Umtrao River in Meghalaya, India.

Economic importance. This species is a potential ornamental fish due to its small size and beautiful coloration.

Conservational status: This species is restricted in the Ranga and Subansiri rivers (Brahmaputra basin). IUCN status: Not assessed (NA).

***Parambassis baculis* (Hamilton 1822)**

Fig. 199.

FIG. 199 *Parambassis baculis*, RGUMF-403, 50.3 mm SL: lateral view.

Chanda baculis Hamilton 1822: 112, 371; type locality: Northeastern Bengal.

Parambassis baculis: Mayanglambam & Vishwanath, 2015: 587 (valid).

Material examined. RGUMF-403, 05 exs., 48.3–50.3 mm SL. Purchased from Doimukh Bazar, Papum Pare district, Arunachal Pradesh.

Vernacular names in Arunachal Pradesh. Not available (in Nyishi, Adi, & Galo).

Diagnosis. Body small, short, compressed, and transparent; jaws short and equal; scales minute, lateral line complete with 80–100 scales. First dorsal with 8 spines and the second dorsal with a spine and 12–13 soft rays; anal with 3 spines and 11–12 soft rays, pectoral with 10–13 soft rays, pelvic fin with a spine and 5 rays.

Coloration. In 10% formalin, body background brownish to yellowish, no markings on the body except a large brownish spot on the nape. Pectoral, pelvic, and anal fins hyaline.

Distribution. Known from India, Bangladesh, and Myanmar. In Arunachal Pradesh: abundantly found in the Subansiri, Dikrong, and Ranga rivers.

Economic importance. This species is a popular ornamental fish due to its small size and coloration.

Conservational status: This fish is wide spread species occurring in the Ganga-Brahmaputra river basins. IUCN status: Least concern (LC).

VIII. ORDER: CICHLIFORMES

The order Cichliformes consist of two families, that is, Cichlidae and Pholidichthyiadae. The members of Cichlidae are characterized by its one nostril each on sides; lateral line interrupted; generally 20–50 scales in lateral lines but number may be exceed 100; generally 7–25 spines and 5–30 soft rays in dorsal fins and 3–15 spines and 4–30 anal-fin soft rays; body shape may differ among the members: moderately deep or compressed body; the body may also be disc shaped or with extremely high sail-like fin or low fin or it can be elongate. While, the members of Pholidichthyiadae are characterized by its one nostril each on sides; scales absent; eel-like body shape; pelvic fin located below or slightly anterior to pectoral base, with one thin spine and 2–3 soft rays, rarely absent; caudal fin rounded and joined with dorsal and anal fins; 66–98 dorsal fin soft rays; 49–81 anal fin soft rays; 15 pectoral fin rays; lower pharyngeals fused into a single bone; septal bone present in interorbital area; larvae with four adhesive attachment glands between the eyes, and 71–101 vertebrae.

24. Family: Cichlidae

87. Genus: *Oreochromis* Günther, 1889

Oreochromis Günther 1889: 70. Type species: *Oreochromis hunter* Günther 1889. Type by monotypy.

Body deep and compressed; mouth terminal, cleft extending below the anterior rim of eye; jaws with 3–5 rows of teeth; spinous portion of dorsal fin longer than the soft part which usually produce into a filament; lateral line interrupted; caudal fin truncate or rounded.

Oreochromis mossambica **(Peters 1852)**
Fig. 200.

FIG. 200 *Oreochromis mossambica*, RGUMF-153, 88.7 mm SL: lateral view.

Chromis mossambicus Peters, 1852: 681; type locality: Mozambique: Tette, Sena, Quellimane, Lumbo, Inbambane, Querimba.

Oreochromis mossambica: Kottelat, 2013: 374 (valid).

Material examined. RGUMF-153, 8 exs., 40–105 mm SL, Poma River at Ramgath, Papum Pare district, Arunachal Pradesh.

Vernacular names in Arunachal Pradesh. *Tilaphia* (in Galo), *Mide* (in Adi).

Diagnosis. Spinous part of dorsal fin longer than soft part with 15–16 spines and 10–12 rays, lateral line interrupted with 30–32 scales, anal fin with 3 spines and 10–11 rays, caudal fin truncate.

Coloration. In live, adult fishes shows watery gray to yellowish, with three or four dark blotches often apparent along flanks. Mature males deep black in color during breeding season. Dorsal fin blackish with a red distal margin; pectoral fin translucent red and caudal fin with a broad red margin.

Distribution. Native in Southeast Africa; introduced widely elsewhere including India. In Arunachal Pradesh: introduced and widely cultured in all foot hill areas.

Economic importance. This species is a popular food fish.

Conservational status: A widely spread species occurring in the multiple basins. IUCN status: Not assessed (NA).

IX. ORDER: BELONIFORMES

Interarcual cartilage which connects the epibranchial bone of the first gill arch with the infrapharyngobranchial of the second gill arch is small or may be absent; the second and third epibranchials are small; lacking both interhyal and dorsal hypohyal; parietal small or may absent; number of principal caudal fin rays in lower lobe is greater than the upper lobe. Presence of a fixed or nonprotrusible upper jaw. About 98 freshwater or brackish water species are known under this order.

25. Family: Belonidae

88. Genus: *Xenentodon* Regan, 1911

Xenentodon Regan 1911: 332. Type species: *Esox cancila* Hamilton 1822. Type by subsequent designation.

Body elongate, subcylindrical and compressed; head and snout sharply pointed; both jaws prolonged into a long beak; large canine teeth alternating with more numerous conical teeth on jaws, palate edentate; a deep median longitudinal groove on the upper surface of head; gill raker absent; dorsal fin without a spine, inserted above anal fin; anal fin with 15–19 rays; caudal-fin truncate.

***Xenentodon cancila* (Hamilton 1822)**
Fig. 201.

FIG. 201 *Xenentodon cancila*, Lateral views of: (A) RGUMF-363, male, 237 mm SL; and (B) RGUMF-364, female, 175 mm SL.

Esox cancila Hamilton, 1822: 213, 380, pl. 27, Fig. 70; type locality: Ponds and smaller rivers of Gangetic provinces, India, and Bangladesh.

Xenentodon cancila: Kottelat, 2013: 298 (valid).

Materials examined. RGUMF-0142, 4 exs., 95.0–168 mm SL, Ranga River at Kimin, Papum Pare district, Arunachal Pradesh (AP). RGUMF-363–364, 02 ex., 170–242 mm SL, Sinkin River at Anpum, Lower Dibang valley district, AP. RGUMF-unregistered specimens: 20 exs., 120–142 mm SL, Dikrong and Ranga rivers in Papum Pare district; Subansiri river in Lower Subansiri district; Siang River at Pasighat, East Siang district; Tengapani River in Namsai district, AP.

Vernacular names in Arunachal Pradesh. *Dangkeng* (in Adi).

Diagnosis. Head and snout sharply pointed. Upper and lower jaws extended into a long "beak" filled with sharp teeth, lower jaw longer than upper. Mature male with a long dorsal fleshy hump just behind head. Dorsal fin without a spine, inserted above anal fin. Dorsal and anal fins about equal in size and almost opposite to each other. Dorsal fin with ii,14 rays and anal fin with ii,15 rays, pectoral fins short with i,11 rays, pelvic fins with I,5 soft rays. Caudal-fin rounded or truncate with 15 principle rays and gill rakers absent.

Coloration. In live, body greenish to light yellowish above and silvery below. A dark silvery lateral band extending on flank of the body. Males tend to possess black edges on the dorsal and anal fins which are absent in females. Opercular region and breast become yellowish gold.

Distribution. Known from India, Pakistan, Bangladesh, Nepal, Sri Lanka, Myanmar, and Thailand. In Arunachal Pradesh: see material examined section.

Remarks. Male have a fleshy hump at dorsum. Pectoral fin with 12 rays (1 simple and 11 branched rays) in all the specimens collected from AP (vs. 9 pectoral fin rays in Hamilton, 1822).

Economic importance. Generally considered as an ornamental fish, besides a food fish.

Conservational status: A widely spread species occurring in the multiple basin of Southeast Asia. IUCN status: Least concern (LC).

X. ORDER: SYNBRANCHIFORMES

The order is characterized by its elongated body; absence of pelvic fins; gill openings confined to lower half of body; enlarged ectopterygoid; reduced mesopterygoid or may absent; nonprotrusible premaxillary and lacking the ascending process; palatine sutured along the posteroventral surface of vomerine shaft, hyohyoidei abductors dorsolaterally expanded, operculum seal to the body wall restricting opercular opening. The order has 3 families and about 117 species.

26. Family: Synbranchidae

89. Genus: *Monopterus* Lacepède, 1800

Monopterus Lacepède 1800: 138. Type species: Masc. *Monopterus javanensis* Lacepède 1800. Type by monotypy.

Body elongated and eel like; head conspicuous and short; snout rounded; mouth terminal and wide; eyes superior, located on anterior half of head; jaws with bands of villiform teeth in jaws and palate; gill openings triangular or crescentic and attached to isthmus internally; suprabranchial accessory respiratory chamber may be present; dorsal and anal fin present in the form of ridges; caudal fin absent; scales embedded in skin or absent or minute; lateral line present.

***Monopterus cuchia* (Hamilton 1822)**
Fig. 202.

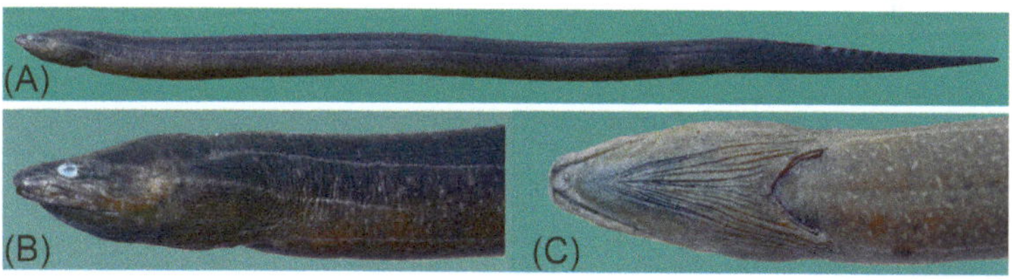

FIG. 202 *Monopterus cuchia*, RGUMF-188, 470 mm SL: (A) lateral view, (B) lateral view of head, and (C) ventral view of head.

Unibranchiapertura cuchia Hamilton, 1822: 16, 363, pl. 16, Fig. 4; type locality: Southeastern Bengal, India.

Monopterus cuchia: Britz et al., 2016: 321 (valid).

Materials examined. RGUMF-188, 4 exs., 300–550 mm SL, Dikrong River at Doimukh, Papum Pare district, Arunachal Pradesh (AP). RGUMF-unregistered specimens 4 exs., 317–410.3 mm SL, Tengapani River at Namsai, Namsai district; Lohit River at Tezu, Lohit district, AP.

Vernacular names in Arunachal Pradesh. *Ngobe* (in Adi & Galo).

Diagnosis. Body elongated, eel like, and cylindrical; scales minute and embedded in skin; head conspicuous. Dorsal fin rudimentary; lacking pectoral, pelvic, anal and caudal fins. Lateral line complete and runs straight midlaterally.

Coloration. In live, body dorsally dark green to chestnut brown, and dirty pale red below. Numerous black spot scattered on body. Two parallel yellow pale lines run forward from the shoulder.

Distribution. Known from India, Bangladesh, Nepal, Pakistan, and Myanmar. In Arunachal Pradesh: see material examined.

Economic importance. This species is regarded as a food fish.

Conservational status: A widely spread species occurring in the multiple basin of Southeast Asia. IUCN status: Least concern (LC).

27. Family: Chaudhuriidae

90. Genus: *Pillaia*, Yazdani, 1972

Pillaia Yazdani, 1972: 134. Type species: *Pillaia indica* Yazdani 1972. Type by original designation (also monotypic).

Body elongated and eel like, subcylindrical; scaleless; mouth terminal its gape extending horizontally to anterior rim of orbit; snout conical with indistinct rostral appendage, gill membranes free from isthmus and also from each other; dorsal fin with long 34–36 rays, inserted slightly posterior to middle of body but slightly ahead of vent, posteriorly confluent with caudal; anal fin with 34–36 rays and confluent with caudal; caudal fin with 8–12 rays; lateral line complete.

Pillaia indica **Yazdani 1972**

Fig. 203.

FIG. 203 *Pillaia indica*, LGCFM-unregistered, 60.6 mm SL, lateral views: (A) preserved, and (B) live views. (Photograph courtesy of Dr. Bikram Bakalial.)

Pillaia indica, Yazdani 1972: 134, Fig. 1, type locality: Sumer stream, about 22 km north of Shillong, Khasi and Jaintia Hills, Meghalaya, India, elevation 1069 m.

Pillaia indica Viswanath et al., 2007: 215 (valid).

Material examined. LGCFM-unregistered, 60.6 mm SL, 1 ex., Subansiri River.

Vernacular names in Arunachal Pradesh. *Ngobe* (in Adi).

Diagnosis. Body eel-like, spine and scales absent, caudal fin confluent with dorsal and anal fins; dorsal fin with 34–36 rays; anal fin with 34–36 rays; caudal fin with 8–10 rays; fins spineless.

Coloration. In 10% formalin, brownish back, yellowish below; indistinct V-shaped marking on the flank.

Distribution. Known from northeastern India: West Bengal, Meghalaya, Assam, and Arunachal Pradesh.

Economic importance. Generally considered as an ornamental and table food fish.

Conservational status: A widely spread species occurring in the Ganga-Brahmaputra river basins. However, their population has rarely beenencountered. IUCN status: Endangered (EN).

28. Family: Mastacembelidae

91. Genus: *Macrognathus* Lacepede, 1800

Macrognathus Lacepède 1800: 283. Type species: *Ophidium aculeatum* Bloch 1786. Type by subsequent designation.

Body eel like and compressed; head pointed; snout fleshy and long with a concave roof that bears series of paired tooth plates; anterior nostril margin with 6 finger-like projections; mouth inferior, jaws subequal, minute pointed teeth present on both jaws, palate and vomer; preorbital and preoperculum smooth; gill rakers absent; dorsal fin origin located far behind distal margin of pectoral fin with 14–22 spines and 42–60 rays; anal fin with 3 spines and 42–58 rays. Caudal fin rounded, separated from both dorsal and anal fins.

Key to the species of *Macrognathus*:

1a. Dorsal fin with 16–23 spines and 44–45 soft rays; 3–5 black ocelli at base of dorsal fin..*M. aral*

1b. Dorsal fin with 24–26 spines and 30–42 soft rays; several minute dark spots irregularly scattered on dorsal fin..................................*M. pancalus*

***Macrognathus aral* (Bloch and Schineider 1801)**

Fig. 204.

FIG. 204 *Macrognathus aral*, RGUMF-457, 154 mm SL: lateral view.

Rhynchobdella aral Bloch & Schneider, 1801: 479, pl. 89; type locality: Tranquebar (Tharangambadi) India.

Macrognathus aral: Kottelat, 2013: 311 (valid).

Materials examined. RGUMF-143, 4 exs., 90–120 mm, Ranga River at Kimin, Papum Pare district, Arunachal Pradesh (AP). RGUMF-457, 12 exs., 86.4–154 mm SL, Dikrong & Ranga rivers in Papum Pare district, AP; Subansiri River in Lower Subansiri district; Siang River at Pasighat, East Siang district, AP.

Vernacular names in Arunachal Pradesh. Not available (in Nyishi, Adi, & Galo).

Diagnosis. Body eel like, snout fleshy and long. Upper jaw with 16–23 pairs of rostral tooth plates. Dorsal with 16–23 spines, 44–45 soft rays; anal fin with 3 spines. Rounded caudal fin with 15 rays; dorsal and anal fins distinctly separated from caudal fin. About 3–5 black ocelli at base of dorsal fin.

Coloration. In live, body brownish or greenish, becoming yellowish along the abdomen. 3–5 blackish ocelli at base of dorsal fin.

Distribution. Widely distributed throughout India, Pakistan, Nepal, and Bangladesh. In Arunachal Pradesh: Currently, occurring in the Kameng, Dikrong, Siang, Noadhing, and Buridhing rivers (Nath and Dey, 2000) (see material examined section for additional range of distribution).

Remarks. Dorsal spines of *M. aral* are more numerous than the *M. pancalus.*

Economic importance. Generally consider as an ornamental fish.

Conservational status: A widely spread species occurring in the Ganga-Brahmaputra river basins. IUCN status: Least concern (LC).

Macrognathus pancalus **Hamilton 1822**

Fig. 205.

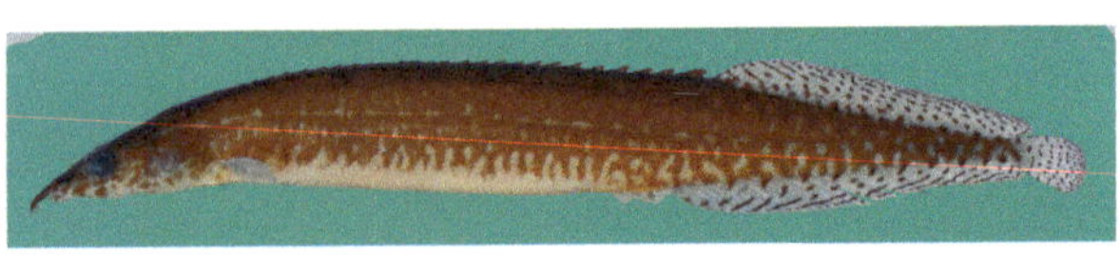

FIG. 205 *Macrognathus pancalus*, RGUMF-144, 110.0 mm SL: lateral view.

Macrognathus pancalus Hamilton, 1822, 30, 364, pl. 22, Fig. 7; type locality: Ganges River drainage, India.

Macrognathus pancalus: Ahmed et al. 2013: 287 (valid).

Materials examined. RGUMF-144, 7 exs., 70.0–115 mm SL, Ranga River at Kimin, Papum Pare district, Arunachal Pradesh (AP). RGUMF-uregistered specimens: 25 exs., 55.8–95.1 mm SL, Dikrong and Ranga rivers in Papum Pare district; Subansiri River in Lower Subansiri district; Sinkin River in Anpum, Lower Dibang Valley district, AP.

Vernacular names in Arunachal Pradesh. Not available (in Nyishi, Adi, & Galo).

Diagnosis. A distinctive streak of longitudinal spots runs along the lateral line from eye to the base of caudal fin. Preopercle with 2–5 spines, preorbital spine strong. Dorsal fin inserted above middle of pectoral fin with 24–26 spines and 30–42 soft rays; anal fin with 3 spines and 31–46 soft rays; lateral line complete; caudal fin separate from dorsal and anal fins.

Coloration. In live, body is greenish olive along back, white marked on the sides below, yellowish on belly, numerous small yellowish white spots on flank. Several minute dark spots irregularly scattered on dorsal, anal, and caudal fins.

Distribution. Widely distributed in the Ganga-Brahmaputra river drainage of India, Bangladesh, and Nepal; and Pakistan (Indus). In Arunachal Pradesh: Kameng, Subansiri, Ranga, Dibang, Lohit, Dikrong, Siang, Noadhing, and Buridhing rivers (Nath and Dey, 2000). Other range of distribution: see material examined.

Remarks. *M. pancalus* is also known as barred spiny eel or Indian spiny eel. It is usually found in beels, ponds, inundated fields with slow moving and shallow rivers. Males are more slender and often smaller than the females.

Economic importance. Generally consider as an ornamental fish.

Conservational status: A widely spread species occurring in the Ganga-Brahmaputra river basins. IUCN status: Least concern (LC).

92. Genus: *Mastacembelus* Scopoli, 1777

Mastacembelus Scopoli (ex Gronow) 1777: 458. Type species: Masc. *Ophidium* [sic] *mastacembelus* Banks & Solander 1794. Type apparently by subsequent monotypy.

Body eel like and compressed; long snout and conical, no tooth plates on the under surface; anterior nostril margin with two finger-like projections; mouth inferior, jaws subequal; minute teeth on jaws and palate; eyes small; dorsal fin inserted above pectoral fin with 24–40 spines and 50–90 soft rays; anal fin with 3 spines and 31–98 soft rays; caudal rounded and usually confluent with dorsal and anal fins; with caudal fin; scales present.

***Mastacembelus armatus* (Lacepede 1800)**
Fig. 206.

FIG. 206 *Mastacembelus armatus*, RGUMF-145, 295 mm SL: lateral view.

Macrognathus armatus Lacepède, 1800: 283, 286; type locality: unknown.
Mastacembelus armatus: Kottelat, 2013: 312 (valid).

Materials examined. RGUMF-145, 5 exs., 120–375 mm SL, Ranga River at Kimin, Papum Pare district, Arunachal Pradesh (AP). RGUMF-unregistered,

8 exs., 102–117 mm SL, Tengapani River in Namsai, Namsai district; and Dikrong and Ranga rivers in Papum Pare district, AP.

Vernacular names in Arunachal Pradesh. Not available (in Nyishi, Adi, & Galo).

Diagnosis. Body elongated and eel like; snout long and conical without transversely striated tooth plates on the under surface; gap of mouth extended posteriorly below posterior nostrils, dorsal and anal fins confluent with caudal fin, dorsal spines 35–36, preopercle with 2 or 3 spines which are usually conspicuous, but often one or more may be embedded in skin; caudal fin with 14–15 rays.

Coloration. In live, body ground color yellowish to dark brown and superimpose with a midlateral blackish zigzag band from eye to caudal-fin base, another thin blackish zigzag lines marking along ventral half of flank, a row of black spots are present along the base of dorsal fin.

Distribution. Known from India, Nepal, Bangladesh, Pakistan, Myanmar, Sri Lanka, Thailand, Cambodia, and South China. In Arunachal Pradesh: Kameng, Subansiri, Siang, Dibang, Lohit, Noadhing, and Buridhing rivers (Nath and Dey, 2000). Additional range of distribution: see material examined.

Remarks. *M. armatus* which is popularly known as tire-track spiny eel or zigzag eel. This species is the largest among the freshwater spiny eels.

Economic importance. It is a popular table fish because of its delicious taste and high nutritional value.

Conservational status: A widely spread species occurring in the multiple basin of Southeast Asia. IUCN status: Least concern (LC).

XI. ORDER: ANABANTIFORMES

Presence of suprabranchial or labyrinth organ; distinctive efferent and afferent circulation to and from the suprabranchial organ; basioccipital with paired articulate processes forming diarthrosis with the upper jaw; the posterior extension of swim bladder extending posteriorly to parhypural; larvae with bilateral pair of oil vesicle used as floating device. The order has 4 families and 207 species.

29. Family: Anabantidae

93. Genus: *Anabas* Cloquet, 1816

Anabas Cloquet (ex Cuvier) 1816: 35. Type species: *Perca scandens* Daldorff 1797. Type by monotypy.

Body oblong; compressed; mouth small, terminal and oblique; first epibranchial arch accommodate a labyrinthine suprabranchial organ for accessory respiration; dorsal fin, inserted above pectoral fin, with 17–18 spines and 8–10 rays; anal fin with 8–10 spines and 9–11 rays; caudal fin rounded; scales ctenoid; lateral line interrupted, the anteriorly upper and posteriorly lower.

***Anabas testudineus* (Bloch 1792)**
Fig. 207.

FIG. 207 *Anabas testudineus*, RGUMF-155, 96.2 mm SL: lateral view.

Anthias testudineus Bloch, 1792: 121, pl. 322; type locality: Japan, actual is southeastern Asia.

Anabas testudineus: Kottelat, 2013: 447 (valid).

Materials examined. RGUMF-155, 8 exs., 80–115 mm SL, Dikrong River at Doimukh, Papum Pare district, Arunachal Pradesh. RGUMF-unregistered, 2 exs., 86.2–106.5 mm SL, purchased from Naharlagun fish market, Papum Pare district, Arunachal Pradesh.

Vernacular names in Arunachal Pradesh. *Koi* (in Nyishi), *Mide* (in Adi).

Diagnosis. Body oblong and laterally compressed, abdomen rounded. Head moderately compressed, snout obtusely pointed. Lower jaw slightly longer than the upper jaw; opercle and preopercle serrated. Lateral line interrupted at 17th scale with 22–29 in lateral series. Dorsal fin single with 17–18 spines and 8–10 rays. Anal fin shorter than the dorsal, with 9–10 spines and 9–10 rays.

Coloration. In live, light Greenish to dark gray on dorsal side and flanks, fading to pale yellow on belly. A distinct blackish spot at base of caudal fin. Fins with pale orange.

Distribution. Known from India, Pakistan, Singapore, Philippines, Bangladesh, and Myanmar. In Arunachal Pradesh: Kameng, Dikrong, Siang, Noadhing, and Buridhing rivers (Nath and Dey, 2000). Other range of distribution: see material examined.

Economic importance. This species is a popular food fish.

Conservational status: A widely spread species occurring in the multiple basin of Southeast Asia. IUCN status: Data deficient (DD).

30. Family: Osphronemidae

Subfamily: Luciocephalinae

94. Genus: *Trichogaster* Bloch & Schneider, 1801

Trichogaster Bloch & Schneider 1801: 164. Type species: Fem. *Trichogaster fasciata* Bloch & Schneider 1801. Type by subsequent designation.

Body short and compressed; mouth terminal, upturned; jaws subequal and protractile; minute villiform teeth on both jaws, absent on palate; opercle entire and ventral margin of preopercle serrated; a suprabranchial organ present, branchial arches toothed; scales ctenoid; lateral line present or absent, interrupted if present; dorsal fin spines 15–18; anal fin spines 15–20; pelvic fin in the form of single filamentously elongated filiform ray; caudal fin slightly emarginate or truncate.

Key to species *Trichogaster*:

1a. Presence of a grayish midlateral stripe on flank .... *Trichogaster chuna*
1b. No stripe on flank... 2
2a. Concave cranio-dorsal profile...
............................*T. lalius*
2b. Convex cranio-dorsal profile...
............................*T. fasciata*

***Trichogaster chuna* (Hamilton 1822)**
Fig. 208.

FIG. 208 *Trichogaster chuna*, RGUMF-445, 32.0mm SL: lateral view.

Trichopodus chuna Hamilton, 1822: 121, 372; type locality: Ganges River, India.

Trichogaster chuna: Ataur-Rahman, 2003: 304 (valid).

Materials examined. RGUMF-445, 08 exs., 24.0–36.0mm SL, stream at Ruksin, East Siang district, Arunachal Pradesh. RGUMF-unregistered, 18 exs., 20.3–28.8mm SL, purchased from Naharlagun fish market in Papum Pare district, AP.

Vernacular names in Arunachal Pradesh. *Colisa* (in Nyishi), *Mide* (in Adi).

Diagnosis. Adult body size small, oblong and compressed. Body dull color with a grayish stripe running from the head to the base of caudal fin. Predorsal with 7–8 scales.

Coloration. In live, body dull greenish, fading toward abdomen, a grayish stripe consisting of black dots running from head to the base of caudal fin. Caudal fin with a black spot at its base.

Distribution. Known from India, Bangladesh, and Pakistan. In Arunachal Pradesh: see material examined.

Economic importance. This species is a popular ornamental fish because of its small size and beautiful coloration.

Conservational status: A widely spread species occurring in the Ganga-Brahmaputra and Indus river basins. IUCN status: Least concern (LC).

***Trichogaster lalius* (Hamilton 1822)**

Fig. 209.

FIG. 209 *Trichogaster lalius*, RGUMF-446, 34.1 SL: lateral view.

Trichopodus lalius Hamilton, 1822: 120, 372; type locality: Gangetic provinces, India.

Trichogaster lalius: Ahmed et al., 2013: 287 (valid).

Materials examined. RGUMF-446, 7 exs., 29.0–37.2 mm SL, stream at Ruksin, East Siang district, Arunachal Pradesh. RGUMF-unregistered, 18 exs., 20.3–28.8 mm SL, Poma and Senkhi rivers and also purchased from Naharlagun fish market in Papum Pare district, Arunachal Pradesh (AP). RGUMF-unregistered, 5 exs., 27.2–28.7 mm SL, Tengapani River in Namsai, Namsai district, AP.

Vernacular names in Arunachal Pradesh. *Colisa* (in Nyishi), *Mide* (in Adi).

Diagnosis. Cranio-dorsal profile somewhat concave, preorbital denticulated, lower limb of preopercle serrated, subopercle entire; soft part of dorsal and anal fin rounded, not produced; 26–28 lateral line scales; caudal fin with 15 rays, and posterior margin rounded.

Coloration. In live, body with alternate scarlet and blue vertical bands, half of each scale being either scarlet or blue color. Dorsal, anal, and caudal fins barred with scarlet markings.

Distribution. Known from India, Bangladesh, and Pakistan. In Arunachal Pradesh: see material examined.

Economic importance. This species is a popular ornamental fish because of its small size and beautiful coloration.

Conservational status: A widely spread species occurring in the Ganga-Brahmaputra and Indus river basins. IUCN status: Least concern (LC).

***Trichogaster fasciata* Bloch & Schneider 1801**
Fig. 210.

FIG. 210 *Trichogaster fasciata*, RGUMF-447, 56.0 mm SL: lateral view.

Polyacanthus fasciatus Bloch & Schneider, 1801: 164; type locality: Tranquebar, India.

Trichogaster fasciatus: Kottelat, 2013: 457 (valid).

Materials examined. RGUMF-157, 11 exs., 52.0–69.0 mm SL, Poma River at Ramgath, Papum Pare district, Arunachal Pradesh (AP). RGUMF-447, 2 exs., 56.0–58.0 mm SL, stream at Ruksin, East Siang district, AP. RGUMF-unregistered, 14 exs., 47.4–65.4 mm SL, Ranga and Senkhi rivers and also purchased from Naharlagun fish market in Papum Pare district, AP.

Vernacular names in Arunachal Pradesh. *Colisa* (in Nyishi), *Mide* (in Adi).

Diagnosis. Body deep, flanks with 14 or more slightly oblique orange bands bordered with blue bands; cleft of mouth small, lips non-papillated and highly protrusible; soft parts of dorsal and anal fins produced; lateral line interrupted midway; fin counts: xiii–xiv,9–10 dorsal rays; xiv–xvi,15–16 anal rays; 1 pelvic rays; and 15 branched caudal rays.

Coloration. In 10% formalin, body brownish to light grayish with 14 or more slightly oblique lateral bands. Dull and lighter along abdomen.

Distribution. Known from India, Nepal, Bangladesh, and Myanmar. In Arunachal Pradesh: Kameng, Dikrong, Siang, Noadhing, and Buridhing rivers (Nath and Dey, 2000). Additional range of distribution: Ranga, Poma, and Senkhi rivers.

Economic importance. This species is a popular food as well as ornamental fish.

Conservational status: A widely spread species occurring throughout multiple basin of South Asia. IUCN status: Least concern (LC).

31. Family: Channidae

95. Genus: *Channa* Scopoli, 1777

Channa Scopoli (ex Gronow) 1777: 459. Type species: *Channa orientalis* Bloch & Schneider 1801. Type by subsequent monotypy.

Body elongated; anteriorly rounded in cross section; head depressed; abdomen rounded; mouth opening wide, usually extended below beyond posterior rim of orbit; lips moderately thick; jaws equal, pointed teeth on jaws and palate; gill openings wide, gill membranes connected beneath the U- or V-shaped isthmus; suprabranchial chamber with bony lamellae as accessory respiratory organ; dorsal inserted above pectoral, no spine, long with 28–55 rays; anal fin inserted nearly at the middle of body, long, with 19–36 rays; caudal fin rounded, not confluent with both dorsal and anal fins; scales small, head with plate like scales; lateral line abruptly curved.

Key to species:

1a. Pelvic fin present.. 2
1b. Pelvic fin absent.. *C. bleheri*
2a. "V"-shaped isthmus...3
2b. "U"-shaped isthmus.. 4
3a. Presence of a distinct ocellus marking on upper side of caudal fin base.. .... *C. marulius*
3b. No ocellus marking on upper side of caudal fin base .... *C. striata*
4a. Presence of transverse bars on pectoral fin, 32–52 dorsal-fin rays 5
4b. No transverse bars on pectoral fin, 28–32 dorsal-fin rays ....*C. punctata*
5a. Numerous blackish spots on flanks 6
5b. No blackish spots on flanks 8
6a. Pectoral fin banded with few (5–6) solid blackish bars........................... 7
6b. Pectoral fin banded with numerous (20–24) narrow dotted blackish bars......................... *C. barca*
7a. 14–15 distinct zigzag blackish transverse bars on caudal fin, total vertebrae 50–51........*C. melanostigma*
7b. No or fewer (6–7) indistinct blackish transverse bars on caudal fin, total vertebrae 44.....*C. stewartii*
8a. Lateral line scales 47 or more... 9
8b. Lateral line scales less than 47 ....*C. gachua*
9a. 45–47 dorsal-fin rays, 28–30 anal-fin rays... .... *C. aurantimaculata*
9b. 36–38 dorsal-fin rays; 25–26 anal-fin rays... .... *C. pomanensis*

***Channa punctata* (Bloch 1793)**
Fig. 211.

FIG. 211 *Channa punctata*, RGUMF-159, 104.5 mm SL: lateral view.

Ophicephalus punctatus, Bloch, 1793: 139, pl. 358; type locality: no locality, not Malabar coast, India.

Channa punctata, Lalhlimpuia et al., 2016: 349 (valid).

Materials examined. RGUMF-159, 9 exs., 90.0–117 mm SL, Sipu at Doji, West Siang district, Arunachal Pradesh. RGUMF-unregistered, 01 ex., 106.5 mm SL, Dikrong River, Papum Pare district, Arunachal Pradesh.

Vernacular names in Arunachal Pradesh. *Taleuk-ngui* (in Nyishi); *Talum/Talu* (in Adi); *Talo* (in Galo).

Diagnosis. "U"-shaped isthmus; cephalic sensory pores open singly, without multiple branches; presence of an enlarged cycloid scale on each side of the ventral surface of the lower jaw; maxilla and premaxillary processes extending to vertical level of beyond the middle of orbit; lower jaw with 3–6 canines behind a single row of villiform teeth; pectoral fin with no transverse bars; pelvic fin longer than the half of pectoral-fin length; 35–40 lateral line scales; 28–32 dorsal-fin rays, and 19–22 anal-fin rays.

Coloration. In live, body brownish to blackish green on sides, whitish yellow on abdomen, numerous black spots on body. Caudal with alternate black bands.

Distribution. Known from India, Afghanistan, Pakistan, Sri Lanka, Nepal, Bangladesh, Myanmar, and China. In Arunachal Pradesh: Subansiri, Siang, Noadhing, and Dikrong rivers (Nath and Dey, 2000). Other range of distribution: see material examined.

Economic importance. A food fish.

Conservational status: A widely spread species occurring in the multiple basins of Southeast Asia. IUCN status: Least concern (LC).

***Channa striata* (Bloch 1793)**
Fig. 212.

FIG. 212 *Channa striata*, RGUMF-425, 328 mm SL: lateral view.

Ophicephalus striatus Bloch, 1793: 141, 359, pl. 359; type locality: India: Tranquebar Tharangambadi.

Channa striata: Kottelat, 2013: 462 (valid).

Materials examined. RGUMF-160, 3 exs., 85.0–115 mm SL, Sipu at Doji, West Siang district, Arunachal Pradesh (AP). RGUMF-425, 1 ex., 328 mm SL, purchase from Naharlagun Bazar, Papum Pare district, AP.

Vernacular names in Arunachal Pradesh. *Taleuk-ngui* (in Nyishi); *Talum*/*Talu* (in Adi); *Talo* (in Galo).

Diagnosis. A species of the "*Chana marulius* group" having the following combination of characters: "V"-shaped isthmus; cephalic sensory pores arranged in groups; no scales on the lower jaw; 18–20 predorsal scales; 55–58 lateral line scales; transverse scale rows 6/1/10½, 32 circumpeduncular scales; fin counts: 42–45 dorsal rays, 15–17 pectoral rays, and 25–28 anal rays.

Coloration. Body dark brown to blackish above, dirty white to yellowish beneath. Flank with chevron-shaped bars, more distinct below the lateral line. Dorsal, anal, and caudal fins dark. No ocellus spot on caudal fin.

Distribution. Widely found in India, Pakistan, Sri Lanka, southern Nepal, possibly Bhutan, Bangladesh, southern China, and all the nations of Southeast Asia. In Arunachal Pradesh: Subansiri, Siang, Noadhing, and Dikrong rivers (Nath and Dey, 2000). Other range of distribution: see material examined.

Economic importance. It is an important food fish.

Conservational status: A widely spread species occurring in the multiple basins of Southeast Asia. IUCN status: Least concern (LC).

***Channa aurantimaculata* Musikasinthorn 2000**

Fig. 213.

FIG. 213 *Channa aurantimaculata*, RGUMF-390, 288 mm SL: (A) lateral, (B) dorsal, and (C) ventral view of head and belly.

Channa aurantimaculata Musikasinthorn, 2000: 27, Figs. 1–3; type locality: market at Dibrugarh town, Dibrugarh, Assam, India.

Channa aurantimaculata: Knight, 2016: 8588 (valid).

Material examined. RGUMF-390, 01 ex., 288 mm SL, Tengapani River in Namsai, Namsai district, Arunachal Pradesh.

Vernacular names in Arunachal Pradesh. *Taleuk-ngui* (in Nyishi); *Talum/Talu* (in Adi); *Talo* (in Galo).

Diagnosis. A species of the "*Channa gachua* species group" having the following unique combination of characters: isthmus "U"-shaped; pectoral fin with a black blotch at the base, followed by about 5 transverse bars; 7–8 large irregular bars on upper part of body; two enlarged cycloid scales present on each side of the ventral surface of the lower jaw; cephalic sensory pores single; lateral line with 51–54 scales; pelvic-fin length less than half the pectoral-fin length; fin counts: dorsal 45–47 rays, anal 28–30 rays, and pectoral 13–16 rays; total vertebrae 51–52.

Coloration. In live, body ground color yellowish to orange, scales on the head and dorsum grayish with thin orange outline, 7–8 broad grayish to black bars descent from the back to side. Pectoral fin orange in color with a black basal blotch, followed by about 5 solid blackish bands.

Distribution. Known from the Brahmaputra river drainage in northeastern India: Assam (Dibrugarh and Tinsukia district) and Arunachal Pradesh (Namsai and Lohit district).

Economic importance. A potential aquarium fish species because of its attractive coloration, besides being a food fish.

Conservational status: This species is restricted in the upper Brahmaputra drainage. IUCN status: Data deficient (DD).

***Channa marulius* (Hamilton 1822)**

Fig. 214.

FIG. 214 *Channa marulius*, RGUMF-389, 135 mm SL: lateral view.

Ophicephalus marulius, Hamilton, 1822: 65, 367, pl. 17, Fig. 19; type locality: River Ganges, India.

Channa marulius: Kottelat, 2013: 461 (valid).

Material examined. RGUMF-389, 03 exs., 135–137 mm SL, Tengapani River at Namsai, Namsai district, Arunachal Pradesh.

Vernacular names in Arunachal Pradesh. *Taleuk-ngui* (in Nyishi); *Talum/Talu* (in Adi); *Talo* (in Galo).

Diagnosis. "V"-shaped isthmus, multiple cephalic sensory pores arranged in groups, no scales on the lower jaw; presence of a distinct ocellus on the upper part of caudal fin and also 5–6 ocelli along body sides; dorsal fin with 49–55 rays; anal fin with 30–36 rays, pectoral fin with 16–18 rays; pelvic-fin with 6 rays, lateral line with 62–70 scales, and predorsal with 16 scales.

Coloration. In live, body color greenish gray above, belly whitish yellow with distinct orange tinge; 5–6 large ocelli (absent in young specimens) on flank which terminates below lateral line; both mature and young specimens bears a blackish ocellus with yellowish outline on the upper half of the caudal-fin base. Young specimens with an orange colored lateral band.

Distribution. Known from India, Sri Lanka, Pakistan, Bangladesh, Nepal, Myanmar, and Thailand to southern China, Laos and Vietnam. In Arunachal Pradesh: see material examined.

Remarks. This fish is the typical species of the "*Marulius* group," characterized by the presence of a very prominent V-shaped sharp isthmus, cephalic sensory pores arranged in groups, and no scales on the lower jaw.

Economic importance. It is regarded as an important food fish in Northeast India.

Conservational status: A widely spread species occurring throughout the Ganga-Brahmaputra river drainage. IUCN status: Least concern (LC).

***Channa stewartii* (Playfair 1867)**

Fig. 215.

FIG. 215 *Channa stewartii*, RGUMF-424, 100 mm SL: lateral view.

Ophiocephalus stewartii, Playfair, 1867: 14; type locality: Cachar, Assam, India.

Channa stewartii: Knight, 2016: 8589 (valid).

Materials examined. RGUMF-424, 2 exs., 100–158.6 mm SL, Singen River, East Siang district, Arunachal Pradesh (AP). MUMF-21, 4 exs., 102.9–113.5 mm SL, Deopani River at Roing, Lower Dibang valley district, AP.

Vernacular names in Arunachal Pradesh. *Taleuk-ngui* (in Nyishi); *Talum/Talu* (in Adi); *Talo* (in Galo).

Diagnosis. A species of the "*Channa gachua* species group" having the following unique combination of characters: U-shaped isthmus; pectoral fin with 4–5 solid black bars; cephalic sensory pores single; the presence of two enlarged cycloid scales on each side of the ventral surface of the lower jaw; small blackish spots scattered on flank; 46–53 lateral line scales; 38–41 dorsal-fin rays; 24–27 anal-fin rays; pelvic fin about one-third as long as pectoral fin; total vertebrae 44.

Coloration. In live, body light pale to shady brownish and shining blue along the dorsal base; distinct circular black spots in most of the scales; pectoral fin with blackish blotches at the base followed by 4–5 bands.

Distribution. Known from the Ganga-Brahmaputra river basin of India, Bangladesh, and Nepal. In Arunachal Pradesh: see material examined section.

Economic importance. This species is regarded as both ornamental and food fish.

Conservational status: A widely spread species occurring throughout the Ganga-Brahmaputra river drainage. IUCN status: Least concern (LC).

***Channa melanostigma* Geetakumari & Vishwanath 2010**
Fig. 216.

FIG. 216 *Channa melanostigma*, RGUMF-437, 115 mm SL: lateral view.

Channa melanostigma Geetakumari & Vishwanath, 2011: 231, Figs. 1 and 2a; type locality: Lohit River, Brahmaputra river drainage, Tezu, Arunachal Pradesh, India.

Channa melanostigma: Knight, 2016: 8589 (valid).

Materials examined. MUMF-Per 39, 134.8 mm SL; India: Arunachal Pradesh: Lohit district, Lohit River, Brahmaputra drainage: Tezu, 27°54′41″N, 96°10′23″E; K. Nebeshwar Sharma, 29.03.2007. RGUMF-437, 01 ex., 115 mm SL, Tengapani River in Namsai, Namsai district, Arunachal Pradesh.

Vernacular names in Arunachal Pradesh. *Taleuk-ngui* (in Nyishi); *Talum/Talu* (in Adi); *Talo* (in Galo).

Diagnosis. Isthmus "U" shaped; flank with 12–14 alternating blackish and pale grayish transverse bars, black spots scattered throughout the flank; 13–14 predorsal with scales, 28 circumpeduncular scales. Dorsal fin with 35–37 simple rays; anal fin with 23–25 simple rays; pectoral fin with i,14–15 rays; pelvic fin with 5 simple rays; and caudal fin rounded with 14 branched rays.

Coloration. In 10% formalin, dorsal surface of body brownish to blackish brow, belly part pale; 12–14 alternating black and pale transverse bars on the flank. Caudal fin with distinct 14–15 blackish bars with irregular intervals. Pectoral fin with 5–6 blackish bars; dorsal and anal fins with whitish distal edges.

Distribution. Known only from the Brahmaputra drainages in Lohit and Namsai district, Arunachal Pradesh.

Remarks. *C. melanostigma* is closely similar to *C. stewartii* in overall body appearance.

Economic importance. This species is regarded as both ornamental and food fish.

Conservational status: This species is restricted in the upper Brahmaputra drainage. IUCN status: Not assessed (NA).

***Channa bleheri* Vierke 1991**
Fig. 217.

FIG. 217 *Channa bleheri*, LGCFM-unregistered, 100mm SL: (A) lateral, and (B) ventral views. (Photograph courtesy of Dr. Bikram Bakalial.)

Channa bleheri Vierke, 1991: 22; type locality: Upper Diburu near Guijam, Assam, India.

Channa bleheri: Knight, 2016: 8588 (valid).

Materials examined. RGUMF-436, 90.4mm SL, Tengapani River at Namsai district, Arunachal Pradesh. LGCFM-unregistered, 100mm SL, 1 ex., Subansiri River.

Vernacular names in Arunachal Pradesh. *Taleuk-ngui* (in Nyishi); *Talum/Talu* (in Adi); *Talo* (in Galo).

Diagnosis. A species of the "*Channa gachua* species group" having the following unique combination of characters: U-shaped isthmus; pelvic fin absent; pectoral fin with 5–7 alternating blackish and yellowish bands; an enlarged cycloid scale present on each side of the ventral surface of the lower jaw; lateral line complete with 45–50 scales.

Coloration. In live, body grayish to brownish on flank, lighter beneath; pectoral fin with 5–7 bars; dorsal and anal fins dull grayish proximally and reddish along the distal margin. Caudal fin exhibit bright orange and blackish markings.

Distribution. Known only from the Brahmaputra river basin in the states of Assam and Arunachal Pradesh (Lohit, Dikrong, and Ranga rivers) in Northeast India.

Remarks. Male tend to grow larger than female.

Economic importance. This species is regarded as both ornamental and food fish.

Conservational status: This species is restricted in the upper Brahmaputra drainage. IUCN status: Near threatened (NT).

Channa barca (Hamilton 1822)
Fig. 218.

FIG. 218 *Channa barca*, RGUMF-388, 230 mm SL: lateral view.

Ophiocephalus barca Hamilton, 1822: 67, 367, pl. 35, Fig. 20; type locality: Brahmaputra River near Goyalpara [Goalpara], Assam, India.

Channa barca: Knight, 2016: 8588.

Material examined. RGUMF-388, 01 ex., 230 mm SL, Tengapani River at Namsai, Namsai district, Arunachal Pradesh.

Vernacular names in Arunachal Pradesh. *Taleuk-ngui* (in Nyishi); *Talum/Talu* (in Adi); *Talo* (in Galo).

Diagnosis. A species of the "*Channa gachua* species group" having the following unique combination of characters: "U"-shaped isthmus; body sides and dorsal fin with several blackish spots, pectoral fin light orange overlay with numerous dotted blackish bars; two enlarged cycloid scale present on each side of the ventral surface of the lower jaw; dorsal fin long with 50–51 rays; anal fin with 33–35 rays; 61–64 lateral line scales; and 15–16 predorsal scales, total vertebrae 55–56.

Coloration. In live, dorsal and flanks covered with numerous blackish spot, pectorals fin light orange or reddish with numerous narrow blackish transverse bands.

Distribution. Known only from the Brahmaputra basin in Assam (Dibrugarh, Tinsukia, Kamrup, and Morigaon districts) and Arunachal Pradesh (Namsai and Lohit district).

Remarks. Reports of this species from Bangladesh and Nepal need verification.

Economic importance. It is also a highly valued and demanded ornamental and food fish.

Conservational status: This species is restricted in the upper Brahmaputra drainage. IUCN status: Data deficient (DD).

Channa gachua (Hamilton 1822)
Fig. 219.

FIG. 219 *Channa gachua*, RGUMF-unregistered, 94.9 mm SL: lateral view.

Ophicephalus gachua Hamilton 1822: 68, 367, pl. 21, Fig. 21; type locality: ponds and ditches of Bengal, India.

Channa gachua: Kottelat, 2015: 48 (valid).

Material examined. RGUMF-unregistered, 01 ex., 94.9 mm SL, Tengapani River at Namsai, Namsai district, Arunachal Pradesh.

Vernacular names in Arunachal Pradesh. *Taleuk-ngui* (in Nyishi); *Talum/Talu* (in Adi); *Talo* (in Galo).

Diagnosis. Isthmus "U" shaped; pectoral fin with about four transverse bands; cephalic sensory pores open singly, without multiple branches; one or two enlarged cycloid scales present on each side of the ventral surface of the lower jaw; pelvic fin length less than half the pectoral-fin length; 39–48 lateral line scales; 32–38 dorsal-fin rays; 15–18 pectoral rays; 22–27 anal rays; caudal fin rounded with 12 rays.

Coloration. In 10% formalin, body blackish dorsally and gradually lighter below. No bars on upper part of body. Dorsal, anal, and caudal fins with white margin. Pectoral fin with about four grayish transverse bands.

Distribution. Known from India, Afghanistan, Indonesia, South and Central Asia. In Arunachal Pradesh: see material examined.

Economic importance. They are known to be used as aquarium fish and as a popular food fish.

Conservational status: A widely spread species naturally breed in streams, ponds, and various wetland water bodies. IUCN status: Least concern (LC).

***Channa pomanensis* Gurumayum & Tamang 2016**
Fig. 220.

FIG. 220 *Channa pomanensis*, ZSI/APRC P-1066, holotype, 111 mm SL: lateral view. (Photograph courtesy of Dr. Shantabala Devi Gurumayum.)

Channa pomanensis Gurumayum & Tamang, 2016: 177, Figs. 1, 2, 3A, 4, 5, type locality: Poma River at Poma (Brahmaputra basin) about 10 km west to capital town, Itanagar, Papum Pare district, Arunachal Pradesh, India, 27°3′33.85″N, 93°30′32.00″E, elevation 240 m (valid).

Material examined. ZSI/APRC P-1066, holotype, 111.6 mm SL, from Poma River at Poma (Brahmaputra basin) Papum Pare district, Arunachal Pradesh, India.

Vernacular names in Arunachal Pradesh. *Taleuk-ngui* (in Nyishi); *Talum/Talu* (in Adi); *Talo* (in Galo).

Diagnosis. Isthmus "U" shaped; 7 oblique bands on body, extending to lateral line; two cycloid scales on either side of lower jaw; predorsal scales 7–8; 47–51 lateral line scales; 36–38 dorsal-fin rays; 25–26 anal-fin rays; total vertebrae 42–45.

Coloration. In 10% formalin, dorsal surface of head and body darker than flank, lateral sides of body bears 7 oblique blackish bands. Dorsal and anal fins dark brown with cream outer margin and faint brown streak medially. Pelvic fin dusky and fin margin hyaline. Caudal fin brown with 14–15 light brown narrow cross bands with orange distal margin.

Distribution. Currently known only from the Poma River in Papum Pare district, Arunachal Pradesh.

Remarks. This species may belong to the "*Channa gachua* species group" as it possess large scales on each side of the ventral surface of the lower jaw and banded pectoral fin.

Economic importance. A potential aquarium fish.

Conservational status: Currently known from Poma River, a water head of Borgong River (Brahmaputra drainage). IUCN status: Not assessed (NA).

32. Family: Nandidae

96. Genus: *Nandus* Valenciennes, 1831

Nandus Valenciennes in Cuvier & Valenciennes 1831: 481. Type species: *Nandus marmoratus* Valenciennes 1831 (=*Coius nandus* Hamilton 1822). Type by monotypy.

Body deep laterally compressed; head large and compressed; snout pointed; eyes large; mouth terminal and protractile, cleft wide, extending below posterior rim of eye; lower law longer; villiform teeth on jaws, palate and tongue except vomer; preopercle serrated; dorsal fin contiguous, posterior soft part higher than anterior spines; pectoral spine weak; 3 anal spines, moderately strong, second spine largest; caudal fin rounded, scales ctenoid, and lateral line interrupted.

***Nandus nandus* (Hamilton 1822)**

Fig. 221.

FIG. 221 *Nandus nandus*, RGUMF-181, 92 mm SL: lateral view.

Coius nandus Hamilton, 1822: 96, 370, pl. 30, Fig. 32; type locality: Ponds of the Gangetic provinces, India.

Nandus nandus: Kottelat, 2013: 366 (valid).

Materials examined. RGUMF-181, 1 exs., 92 mm SL, Poma River at Ramgath, Papum Pare district, Arunachal Pradesh. RGUMF-unregistered,

14 exs., 86.59–98.2 mm SL, Subansiri River in Lower Dibang Valley district; Tengapani River, Namsai district; Rang and Poma rivers in Papum Pare district, AP.

Vernacular names in Arunachal Pradesh. Not available (in Nyishi, Adi, & Galo).

Diagnosis. Mouth superior with deep cleft and very much protractile; maxilla reaching beyond posterior rim of orbit; operculum with a single prominal spine; preopercle serrated in a continuous band; and all fins with blackish marking. Dorsal fin with 12–14 spines and 11–12 soft rays; anal fin with 3 spines and 7–9 rays; and lateral line incomplete with 46–57 scales.

Coloration. In live, body greenish brown with three vertical broad irregular blackish bands. All fins marked with narrow bands.

Distribution. Known from India, Bangladesh, Nepal, and Pakistan. In Arunachal Pradesh: see material examined.

Remarks. This species can change their body color to camouflage against its environmental condition.

Economic importance. They are commercially important for the aquarium trade and are also highly prized local food fish.

Conservational status: A widely distributed species in lowland water bodies (see distribution section). IUCN status: Least concern (LC).

33. Family: Badidae

97. Genus: *Badis* Bleeker, 1853

Badis Bleeker 1853: 106. Type species: *Labrus buchanani* Bleeker 1853 (=*Labrus badis* Hamilton 1822). Type by tautonymy.

Body moderately elongated and compressed; mouth terminal, upturned and slightly protractile; maxilla not extending behind anterior margin of orbit; villiform teeth on jaws, vomer and palatines; opercles with a sharp spine at posterodorsal corner; spinous and soft dorsal contagious, posterior soft part higher than anterior spinous part; 3 spines on anal fin followed by 6–8 soft rays; lateral line pores tubed and interrupted; scales both ctenoid and cycloid; 2–4 dentary foramina; hypobranchial 3-toothed; caudal fin rounded.

Key to species of *Badis*:

1a. Absence of blackish blotch on soft anal fin................................2
1b. Presence of blackish blotch on soft anal fin............................ *B. singenensis*
2a. No conspicuous blotch behind opercle .... *B. assamensis*
2b. A conspicuous blotch behind opercle 3
3a. A small oval blackish blotch present in the mid-base of caudal fin, interorbital width 9.9–15.0 SL ....*B. dibruensis*
3b. Large rectangular black blotch in the base of caudal fin interorbital width 6.5%–8.3% SL .... *B. badis*

***Badis assamensis* Ahl 1937**
Fig. 222.

FIG. 222 *Badis assamensis*, RGUMF-180, 52.0 mm SL: lateral view.

Badis badis assamensis Ahl 1937: 118; type locality: Rivulet falling in the Dibru River, 3 km north of Digholtarang, southeastern Dubrugarh, Brahmaputra River drainage, Assam, India.

Badis assamensi: Kullander and Britz, 2002: 3334 (valid).

Material examined. RGUMF-180, 2 exs., 49.0–52.0 mm SL, Dibang River in Lower Dibang Valley district, Arunachal Pradesh.

Vernacular names in Arunachal Pradesh. *Ngagae* (in Nyishi), *Mide* (in Adi).

Diagnosis. A prominent black blotch posterodorsally on opercle and two rows of irregular blackish blotches on flanks. Dorsal with 16–17 spines and 9–11 dorsal soft rays; and anal with 7–8 soft rays. Lateral row scales 28–29; circumpeduncular scale rows: 9 above, 9 below lateral lines. Gill raker 7–9 on the first branchial arch.

Coloration. In 10% formalin, blackish brown spot posterodorsally on opercle, at the base of opercular spine. Body side with 10 vertical bars expressed chiefly as small, contrasting black spots arranged in two principal horizontal rows, one row along upper lateral line, second row along middle of side. One dark spot dorsally and another ventrally on caudal fin base. Dorsal fin dark brown with narrow hyaline margin. Anal fin dusky and pelvic fin also dusky with white margin.

Distribution. Known from the upper reaches of the Brahmaputra River in Assam (Dibru River) and Arunachal Pradesh (Dibang River) India.

Remarks. *B. assamensis* is the largest species in the genus, attaining nearly 68 mm SL.

Economic importance. This species is a popular ornamental fish due to its unique coloration.

Conservational status: This species is likely to have a restricted distribution in the upper Brahmaputra river drainage. Data deficient (DD).

***Badis badis* (Hamilton 1822)**
Fig. 223.

FIG. 223 *Badis badis*, RGUMF-149, 35.8 mm SL: lateral view.

Labrus badis Hamilton, 1822: 70, 368, pl. 25, Fig. 23; type locality: Shore of Tumapao River close to Duma village (Ganges River drainage) north-northeast of Calcutta, West Bengal, India.

Badis badis: Kullander and Britz, 2002: 305 (valid).

Material examined. RGUMF-149, 5 exs., 23.0–40.0 mm SL, Mebang River at Boleng, East Siang district, Arunachal Pradesh.

Vernacular names in Arunachal Pradesh. *Ngagae* (in Nyishi), *Mide* (in Adi).

Diagnosis. Body with a prominent blackish blotch posterodorsally behind the opercle and two rows of irregular blackish blotches along flank; a series of prominent black blotches along dorsal-fin base and a series of black blotches along middle of dorsal fin; indistinct bars on side and lacking of the caudal blotch; 27–28 lateral scales row; 19–20 cicumpenduncular scales; 7–9 gill rakers on the first branchial arch; dorsal with 15–17 spines and 7–10 soft rays; anal with 6–8 soft rays; and pectoral with 12 rays soft rays.

Coloration. In 10% formalin, body brownish to black markings. Prominent dark brown cleithral blotch, 5–6 broad vertical dark bars across flank and caudal peduncle. Black bar on caudal peduncle usually well separated from bars between dorsal and anal fin bases. Dorsal fin dusky to grayish, anal fin dusky basally and pelvic fin dark gray.

Distribution. Known from the Ganga-Brahmaputra drainages of Indian, Bangladesh, Bhutan, and Nepal. In Arunachal Pradesh: Mebang River.

Remarks. Males are also more colorful than females (which are without red or blue colors on their flanks), making them the more popular sex for importation.

Economic importance. This species is a popular ornamental fish because of its unique coloration.

Conservational status: A widely spread species occurring throughout the Ganga-Brahmaputra river drainage. IUCN status: Least concern (LC).

Badis singenensis Geetakumari & Kadu 2011
Fig. 224.

FIG. 224 *Badis siengenensis*, RGUMF-218, paratype, 36.5 mm SL: lateral view.

Badis singenensis Geetakumari & Kadu, 2011: 2086, Image 1; type locality: Saku-Kadu Village, Singen River, Brahmaputra drainage, Arunachal Pradesh, India.

Badis triocellus Khynriam & Sen, 2013: 65, pl. I; type locality: Subansiri River below dam site, Lower Subansiri district, Arunachal Pradesh, India.

Badis singenensis: Basumatary et al., 2016: 152 (valid).

Materials examined. RGUMF-218–225, paratypes, 26 exs., 27.0–37.0 mm SL, Singen River, Saku-Kadu Village, Brahmaputra drainage, East Siang district of Arunachal Pradesh (AP). RGUMF-unregistered specimens: 15 exs., 24.2–31.6 mm SL, Siang River at Pasighat, East Siang district; Subansiri river in Lower Subansiri district, AP. RGUMF-421, 9 exs., 26.5–41.5 mm, Lohit River, Timaighat, Lohit district, AP. RGUMF-151, 3 exs., 35.0–52.0 mm SL, Dikrong River at Sagalee, Papum Pare district, AP.

Vernacular names in Arunachal Pradesh. *Ngagae* (in Nyishi), *Mide* (in Adi).

Diagnosis. A prominent blackish blotch posterodorsally on opercle, at the base of opercle spine, 3 distinct dark blotches at dorsal-fin base, first blotch behind third spine, second behind sixth dorsal spine and third behind the fifth and sixth soft dorsal ray. Another distinct darker blotch at the base of anal fin behind the fifth soft anal-fin ray. Scales in lateral row 28–32, 9 circumpeduncular scale rows above, 9 below lateral line, totalling 19–20. Dorsal with 15–17 spines and 7–9 soft rays; pectoral with 13–14 soft rays; and anal with 5–8 soft rays.

Coloration. In 10% formalin, body with 5–6 broad blackish bars. Black bar on caudal peduncle well separated from the preceding bar between the posterior ends of dorsal- and anal-fin bases. Dark blotches on dorsal-fin base and soft anal fins prominent. Dorsal fin grayish, with white distal margin.

Distribution. Known only from Arunachal Pradesh, India: see material examined.

Economic importance. This species is a popular ornamental fish due to its unique coloration.

Conservational status: This species is widely distributed in multiple drainages in Arunachal Pradesh. IUCN status: Not assessed (NA).

***Badis dibruensis* Geetakumari & Vishwanath 2010**
Fig. 225.

FIG. 225 *Badis dibruensis*, RGUMF423, 34.9 mm SL: lateral view.

Badis dibruensis, Geetakumari & Vishwanath 2010: 645, Figs. 1 and 2, type locality: Dibru River, 27°32′09.98″N, 94°58′02.31″E, Brahmaputra river drainage, Assam, India.

Badis dibruensis Basumatary et al., 2016: 152 (valid).

Materials examined. RGUMF-423, 01 ex., 34.9mm SL, Dikrong River at Doimukh, Papum Pare district, Arunachal Pradesh. RGUMF-unregistered, 2 exs., 32.3–33.8mm SL, Tengapani river in Namsai, Namsai district, Arunachal Pradesh (AP).

Vernacular names in Arunachal Pradesh. *Ngagae* (in Nyishi), *Mide* (in Adi).

Diagnosis. A conspicuous dark blotch present over the superficial part of cleithrum above the pectoral-fin base, no vertical bars or markings along flank, a small oval-shaped black blotch on the middle of caudal-fin base, two predorsal bones present; dorsal fin with 15–17 spines and 8–9 soft rays; eyes large with diameter 7.6%–9.4% SL and widely apart with interorbital distance 9.9%–15.0% SL.

Coloration. In 10% formalin, overall body blackish brown to light yellowish. A distinct dark blotch present over the superficial part of cleithrum and another small oval-shaped blackish blotch on the middle of caudal-fin base. No vertical markings along flank. Dorsal and pelvic fins grayish.

Distribution. Know from the Brahmaputra drainage in Assam and Arunachal Pradesh.

Remarks. This species is recorded here for the first time from Arunachal Pradesh.

Economic importance. This fish has high ornamental value.

Conservational status: This species is widely distributed Brahmaputra river drainage. IUCN status: Data deficient (DD).

XII. ORDER: SYNGNATHIFORMES

Body elongated; mouth small, located at the end of tubular snout (except in *Bulbonaricus*); pelvic fin inserted at abdominal region or absent; upper jaw not protractile; lachrymal usually present, other bones in the circumorbital series often absent; no ribs; anterior third to sixth vertebrae elongate; aglomerular kidney present in some forms. The order has 8 families and 338 species.

34. Family: Syngnathidae

98. Genus: *Microphis* Kaup, 1853

Microphis Kaup, 1853: 234. Type species: *Syngnathus cunalcus* Hamilton 1822. Type by subsequent designation.

Body elongated; covered with many distinct ridges; head long; mouth terminal and narrow; jaws produced into a spatulated beak; barbels absent; eyes large, superior, placed in posterior half of head; dorsal fin inserted opposite to vent, long with about 30 rays, caudal fin fan shaped, middle rays prolonged; males with longitudinal brood pouch on abdomen.

***Microphis deocata* (Hamilton 1822)**

Fig. 226.

FIG. 226 *Microphis deocata*, RGUMF-360, 158.2 mm SL: (A) lateral view, and (B) ventral view showing egg pouch of male.

Syngnathus deocata Hamilton, 1822: 14, 363; type locality: Tista, Kuwarlayi and other rivers of Puraniya or Mithila in northern Bengal or Bahar, India.

Microphis deocata: Kuiter, 2009: 275 (valid).

Material examined. RGUMF-360, 02 exs., 145-158 mm SL, Sinkin River at Anpum (a tributary of Siang River) Lower Dibang Valley district, Arunachal Pradesh.

Vernacular names in Arunachal Pradesh. Not available (in Nyishi, Adi, & Galo).

Diagnosis. Body elongated; mouth terminal; jaws produced into a beak; barbels and pelvic fin absent; mature females with "Y"-shaped markings along the lower half of the body, males with a brood pouch on the belly; the dorsal fin with 13 rays, the pectoral fins with 15 rays, and oval caudal fin with 7 rays.

Coloration. In 10% formalin, body brownish, sides of the belly below the lateral lines, in mature females, with "Y"-shaped markings.

Distribution. Known from the Ganga-Brahmaputra river drainages in India and Bangladesh. In Arunachal Pradesh: see material examined.

Remarks. A successful spent male carries the fertilized eggs in their brood pouch on the belly.

Economic importance. This fish has high ornamental value.

Conservational status: This species is restricted to the Brahmaputra river drainage. IUCN status: Near threatened (NT).

XIII. ORDER: TETRAODONTIFORMES

Parietals, nasals, and infraorbitals absent, and ribs usually absent; posttemporal, when present, simple and fused with pterotic; no extrascapulars; hyomandibular and palatine tightly attached to skull; gill opening restricted and placed anteriorly to pectoral fin base; maxilla often joined or fused with premaxilla; scales usually modified as spines, shields, or plates; multiple lateral line may present or totally absent; swim bladder present (except in molids); no anal fin spine; pelvic fin with 1 spine followed by 2 soft rays; caudal fin with 12 principal rays or fewer in some forms; 21 vertebrae or fewer in some forms (e.g., Tetraodontidae, Diodontidae, and Monacanthidae).

35. Family: Tetraodontidae

99. Genus: *Leiodon* Swainson, 1839

Leiodon Swainson, 1839: 194. Type species: *Tetrodon marmoratus* Swainson 1839. Type by monotypy.

Body short with broad back; head oval, snout blunt; mouth terminal, transverse and small; jaws equal, lips thick and fleshy; eyes large and superior and located on posterior half of head, dorsal fin short, without a spine and inserted conspicuously ahead of anal fin; pelvic fin absent; caudal fin rounded or emarginate or truncate; body surface partly with fine dermal spines or the spines totally absent.

***Leiodon cutcutia* (Hamilton 1822)**
Fig. 227.

FIG. 227 *Leiodon cutcutia*, RGUMF-162, 75.0 mm SL: lateral view.

Tetrodon cutcutia Hamilton, 1822: 362; type locality: India: everywhere watered by the Ganges and its branches.

Leiodon cutcutia: Kottelat, 2013: 479 (valid).

Materials examined. RGUMF-162, 3 exs., 52.0–85.0 mm SL, Dikrong River at Doimukh, Papum Pare district, Arunachal Pradesh (AP). RGUMF-unregistered specimens: 12 exs., 42.6–78.2 mm SL, Noadhing River at Lohit district, Siang River at East Siang district, Subansiri River and Sinkin River at Lower Dibang Valley district, AP.

Vernacular names in Arunachal Pradesh. *Pacha* (in Nyishi).

Diagnosis. Body short and oval. Short tube-like nostrils with terminal opening. Scales, lateral line, and pelvic fin are absent. Gill opening restricted to a slit in front of the pectoral fin. Branchiostegal area covered by thick skin. Oesophagus inflatable with air. Dorsal fin short with ii,7 rays, and inserted slightly ahead of anal fin. Pectoral fin with i,19 rays. Anal fin with ii,6 rays.

Coloration. In live, body olive green dorsally with dark reticulated markings. A dark occelus on sides. Caudal fin dusky, other fins yellowish gray to olive green.

Distribution. Known widely distributed in India, Bangladesh, Sri Lanka, Myanmar, and Malay archipelago. In Arunachal Pradesh: see materials examined.

Remarks. This species is generally regarded as a poisonous fish.

Economic importance. This ornamental fish species has high market value.

Conservational status: A widely spread common fish species occurring in multiple river basins in the southeast Asia. IUCN status: Not assessed (NA).

References

Ahl E. Neue Subwasserfische aus dem Indischen und Malaiischen Gebiet. Zool Anz 1937;117(5/6):113–9.

Ahmed AT, Rahman MM, Mandal S. Biodiversity of hillstream fishes in Bangladesh. Zootaxa 2013;3700(2):283–92.

Anganthoibi N, Vishwanath W. *Pseudecheneis koladynae*, a new sisorid catfish from Mizoram, India (Teleostei: Siluriformes). Ichthyol Explor Freshw 2010;21(3):199–204.

Anganthoibi N, Vishwanath W. *Glyptothorax pantherinus*, a new species of catfish (Teleostei: Sisoridae) from the Noa Dehing River, Arunachal Pradesh, India. Ichthyol Res 2013;60(2):172–7.

Arunachalam M, Nandagopal S. A new species of the genus *Garra* Hamilton, (Cypriniformes: Cyprinidae) from Nethravathi River, Western Ghats, India. Species 2014;10(24):43–57.

Arunachalam M, Nandagopal S, Mayden RL. Morphological diagnoses of *Garra* (Cypriniformes: Cyprinidae) from North-Eastern India with four new species description from Brahmaputra River. J Fish Aquac 2013;4(3):121–38.

Arunachalam M, Raja M, Punniyam M, Mayden RL. New species of *Aborichthys* (Cypriniformes: Balitoridae) from Arunachal Pradesh, India. Species 2014;7(18):33–47.

Arunkumar L. *Glyptothorax pasighatensis*, a new species of catfish (Teleostei: Sisoridae) from Arunachal Pradesh, northeastern India. Int J Pure Appl Zool 2016;4(2):179–85.

Arunkumar L, Basudha C. *Tor barakae*, a new species of mahseer fish (Cyprinidae: Cyprininae) from Manipur, India. Aquaculture 2003;4(2):271–6.

Ataur-Rahman AK. Freshwater fishes of Bangladesh. 2nd ed. Dhaka: Zool Soc Bangl; 2003.

Bagra K, Kadu K, Sharma KN, Laskar BA, Sarkar UK, Das DN. Ichthyological survey and review of the checklist of fish fauna of Arunachal Pradesh, India. Checklist 2009;5:330–50.

Bănărescu P, Nalbant TT, Goren M. The noemacheiline loaches from Israel (Pisces: Cobitidae: Noemacheilinae). Isr J Zool 1982;31:1–25. 2 pls.

Bănărescu PM, Herzig SB. A revision of the species of the *Cyprinion macrostomus*-group (Pisces: Cyprinidae). Ann Naturhist Mus Wien 1995;97B:411–20.

Barman RP. A new freshwater fish of the genus *Danio* Hamilton (Pisces: Cyprinidae) from Assam, India, with the key to the identification of the Indian species of the subgenus *Danio*. Bull Zool Sur India 1984;6(1–3):163–5.

Barman RP. A new cobitid fish of the genus *Aborichthys* Chaudhuri (Pisces: Cobitidae) from India. J Bombay Nat Hist Soc 1985;81(pt. 3):680–3.

Basumatary S, Choudhury H, Baishya RA, Sarma D, Vishwanath W. *Badis pancharatnaensis*, a new percoid fish species from Brahmaputra River drainage, Assam, India (Teleostei: Badidae). Vert Zool 2016;66(2):151–6.

Bleeker P. Bijdrage tot de kennis der ichthyologische fauna van Midden- en Oost-Java, met beschrijving van eenige nieuwe species. Verhandelingen van het Bataviaasch Genootschap van Kunsten en Wetenschappen 1849;23(12):1–23.

Bleeker P. Nieuwe bijdrage tot de kennis der ichthyologische fauna van Borneo met beschrijving van eenige nieuwe soorten van zoetwatervisschen. Natuurkundig Tijdschrift voor Nederlandsch Indie 1851;1(3):259–75.

Bleeker P. Nalezingen op de ichthyologische fauna van Bengalen en Hindostan. Verhandelingen van het Bataviaasch Genootschap van Kunsten en Wetenschappen 1853;.

Bleeker P. De visschen van den Indischen Archipel beschreven en toegelicht. Deel I. Siluri. Acta Societatis Regiae Scientiarum Indo-Neêrlandicae 1858;4(art. 2). i–xii + 1–370.

Bleeker P. Enumeratio specierum piscium hucusque in Archipelago indico observatarum, adjectis habitationibus citationibusque, ubi descriptiones earum recentiores reperiuntur, nec non speciebus Musei Bleekeriani Bengalensibus, Japonicis, Capensibus Tasmanicisque. Acta Societatis Regiae Scientiarum Indo-Neêrlandicae 1859;6(art. 3). i–xxxvi + 1–276.

Bleeker P. Conspectus systematis Cyprinorum. Natuurkundig Tijdschrift voor Nederlandsch Indie 1860;20(3):421–41.

Bleeker P. Notice sur les genres Parasilurus, Eutropiichthys, Pseudeutropius, et Pseudopangasius. Verslagen en Mededeelingen der Koninklijke Akademie van Wetenschappen. Afdeling Natuurkunde 1862;14:390–9.

Bleeker P. Sur les genres de la famille des Cobitioïdes. Verslagen en Mededeelingen der Koninklijke Akademie van Wetenschappen. Afdeling Natuurkunde 1863;15:32–44.

Bleeker P. Révision des espèces d'Ambassis et de Parambassis de l'Inde archipélagique. Natuurkundige Verhandelingen van de Hollandsche Maatschappij der Wetenschappen te Haarlem 1874;2(2):83–106.

Bloch ME. Naturgeschichte der ausländischen Fische. Berlin. 1792.

Bloch ME. Naturgeschichte der ausländischen Fische. Berlin. 1793.

Bloch ME. Naturgeschichte der ausländischen Fische. Berlin. vol. 8; 1794. i–iv + 1–174.

Bloch ME, Schneider JG, Blochii ME. Systema Ichthyologiae Iconibus cx Ilustratum. Post obitum auctoris opus inchoatum absolvit, correxit, interpolavit Jo. Gottlob Schneider, Saxo. Berolini. Sumtibus Auctoris Impressum et Bibliopolio Sanderiano Commissum; 1801. i–lx + 1–584.

Blyth E. Report of Curator, Zoological Department, for May, 1858. J Asiat Soc Bengal 1858;27(3):267–90.

Blyth E. Report on some fishes received chiefly from the Sitang River and its tributary streams, Tenasserim provinces. J Asiat Soc Bengal 1860;29(2):138–74.

Bornbusch AH. Monophyly of the catfish family Siluridae (Teleostei: Siluriformes), with a critique of previous hypotheses of the family's relationships. Zool J Linnean Soc 1991;101(2):105–20.

Britz R, Doherty-Bone TM, Kouete MT, Sykes D, Gower DJ. *Monopterus luticolus*, a new species of swamp eel from Cameroon (Teleostei: Synbranchidae). Ichthyol Explor Freshw 2016;27(4):309–23.

Chaudhuri BL. XXXV. Descriptions of some new species of freshwater fishes from North India. Rec Indian Mus (Calcutta) 1912;7(5):437–44.

Chaudhuri BL. Zoological results of the Abor Expedition, 1911–12. XVIII. Fish. Rec Indian Mus (Calcutta) 1913;243–57.

Chen XY, Qin T, Chen ZY. *Oreoglanis hponkanensis*, a new sisorid catfish from North Myanmar (Actinopterygii, Sisoridae). ZooKeys 2017;646:95–108.

Chopra BN. Handbook of Indian Fisheries prepared for the third meeting of the Indo-Pacific Fisheries Council Madras. Ministry of Agriculture New Delhi, Govt. of India; 1951.

Cloquet H. Pisces accounts. Dictionnaire des sciences naturelles 1816;2(Suppl):35–6.

Cuvier G. Le Règne Animal distribué d'après son organisation pour servir de base à l'histoire naturelle des animaux et d'introduction à l'anatomie comparée. In: Les reptiles, les poissons, les mollusques et les annelids. 1816.

Cuvier G, Valenciennes A. Histoire naturelle des poissons. Tome septième. Livre septième. Des Squamipennes. Livre huitième. Des poissons à pharyngiens labyrinthiformes. vol. 7. Paris: F. G. Levrault; 1831. i–xxix + 1–531.

Cuvier G, Valenciennes A. Histoire naturelle des poissons. Tome dix–septième. Suite du livre dix–huitième. Cyprinoïdes. vol. 17. 1844. i–xxiii + 1–497.

Darshan A, Anganthoibi S, Vishwanath W. Redescription of the striped catfish *Mystus carcio* (Hamilton) (Siluriformes: Bagridae). Zootaxa 2010;2475:48–54.

Darshan A, Anganthoibi N, Vishwanath W. *Batasio convexirostrum*, a new species of catfish (Teleostei: Bagridae) from Koladyne basin, India. Zootaxa 2011;2901:52–8.

Darshan A, Dutta R, Abujam S, Das DN. First record of Batasio Spilurus Ng from the Siang River of Arunachal Pradesh, Northeastern India (Teleostei: Bagridae). J FisheriesSciences.com 2017;11(2):021–7.

Darshan A, Mahanta PC, Bara A, Kumar P. Redescription of the striped catfish *Mystus tengara* (Hamilton, 1822) (Siluriformes: Bagridae), India. J Threat Taxa 2013;5(1):3536–41.

Darshan A, Dutta R, Kachari A, Gogoi B, Aran K, Das DN. A new species of glyptosternine catfish (Siluriformes: Sisoridae) from Yomgo River, Arunachal Pradesh, India. Aqua Int J Ichthyol 2014;20(2):73–80.

Darshan A, Dutta R, Kachari A, Gogoi B, Das DN. *Glyptothorax mibangi*, a new species of catfish (Teleostei: Sisoridae) from the Tisa River, Arunachal Pradesh, northeast India. Zootaxa 2015;3962(1):114–22.

Darshan A, Kachari A, Dutta R, Ganguly A, Das DN. *Amblyceps waikhomi*, a new species of catfish (Siluriformes: Amblycipitidae) from the Brahmaputra drainage of Arunachal Pradesh, India. PLoS ONE 2016;11(2):1–10. e0147282.

Das DN, Darshan A. Physoschistura harkishorei, a new species of loach from Arunachal Pradesh, North-Eastern India (Teleostei: Nemacheilidae). Zootaxa 2017;4337(3):403–12.

Datta AK, Barman RP. On a new species of *Noemacheilus* (Pisces: Cobitidae) from Arunachal Pradesh, India. Bull Zool Surv India 1984;6(1–3):275–7.

Datta AK, Barman RP, Jayaram KC. On a new species of *Kryptopterus* (Pisces: Siluroidea, family: Siluridae) from Namdapha Wildlife Sanctuary, Arunachal Pradesh, India. Bull Zool Surv India 1987;8(1–3):29–31.

Day F. Notes on the genus *Hara*. J Asiat Soc Bengal 1870;39(1 (pt 2)):37–40.

Day F. On the fishes of the Andaman Islands. Proc Sci Meet Zool Soc Lond 1871;1870(3):677–705.

Day F. The fishes of India: being a natural history of the fishes known to inhabit the seas and freshwaters of India, Burma and Ceylon (Parts 1–4). London: Berdnard Quaritch; 1875–1878. p. 553–779.

Day F. The fishes of India; being a natural history of the fishes known to inhabit the seas and fresh waters of India, Burma, and Ceylon (Part 3). London: Berdnard Quaritch; 1877369–552.

Day F. The fishes of India; being a natural history of the fishes known to inhabit the seas and fresh waters of India, Burma, and Ceylon (Part 4). London: Berdnard Quaritch; 1878. i–xx + 553–779.

Dutta Choudhary S, Sen N. On a collection of fish from Arunachal Pradesh with some new records. Newsl Zool Surv India 1977;3(4):217–23.

Endruweit MA. New *Schistura* from the Salween basin in western Yunnan (Teleostei: Nemacheilidae). Zootaxa 2017;4243(2):394–400.

Eschmeyer WN, editor. Catalog of fishes; 2017. Updated database version of 4 April 2017. Catalog databases of CAS cited in FishBase (website).

Fang F. Phylogenetic analysis of the Asian cyprinid genus *Danio* (Telostei, Cyprinidae). Copeia 2003;3(4):714–28.

Ferraris Jr CJ. Checklist of catfishes, recent and fossil (Osteichthyes: Siluriformes), and catalogue of siluriform primary types. Zootaxa 2007;1418:1–628.

Forsskål PS. Descriptiones animalium avium, amphibiorum, piscium, insectorum, vermium; quae in itinere orientali observavit. Post mortem auctoris edidit Carsten Niebuhr. Hauniae; 1775. 1–20 + i-xxxiv + 1–164, map.

Fowler HW. Notes on clupeoid fishes. Proc Acad Nat Sci Phila 1911;63:204–21.

Fowler HW. Descriptions of new fishes obtained 1907 to 1910, chiefly in the Philippine Islands and adjacent seas. Proc Acad Nat Sci Phila 1934a;85:233–367.

Fowler HW. Zoological results of the Third de Schauensee Siamese Expedition, part I. – Fishes. Proc Acad Natl Sci Phila 1934b;86:67–163. pl. 12.

Fricke R. Annotated checklist of the marine and estuarine fishes of Germany, with remarks of their taxonomic identity. Stuttgarter Beiträge zur Naturkunde, Serie A (Biologie) 1999;587:1–67.

Geetakumari K. *Parambassis bistigmata*, a new species of glassperch from North–Eastern India (Teleostei: Ambassidae). Zootaxa 2012;3317:59–64.

Geetakumari K, Kadu K. *Badis singenensis*, a new fish species (Teleostei: Badidae) from Singen River, Arunachal Pradesh, northeastern India. J Threat Taxa 2011;3(9):2085–9.

Geetakumari K, Vishwanath W. *Channa melanostigma*, a new species of freshwater snakehead from North–East India (Teleostei: Channidae). J Bombay Nat Hist Soc 2011;107(3):231–5.

Gill TN. Description of a new generic form of Gobinae from the Amazon River. Ann Lyceum Nat Hist New York 1859;7(1–3):45–8.

Grant S. A new subgenus of *Acanthocobitis* Peters, 1861 (Teleostei: Nemacheilidae). Ichthyofile 2007;2:1–9.

Gray JE. Illustrations of Indian zoology; chiefly selected from the collection of Major General Hardwicke, F.R.S. Paris/London: Treuttel, Wurtz, Treuttel, Jun. and Richter/Parbury, Allen and Co; 1830–1835.

Gray JE. Description of twelve new genera of fish, discovered by Gen. Hardwicke, in India, the greater part in the British Museum. Zool Misc 1831;7–9.

Greenwood PH. A review of the family Centropomidae (Pisces: Perciformes). Bull Br Mus Nat Hist 1976;29(1):1–81.

Günther A. Catalogue of the fishes in the British Museum. Catalogue of the Physostomi, containing the families Heteropygii, Cyprinidae, Gonorhynchidae, Hyodontidae, Osteoglossidae, Clupeidae, [thru]. Halosauridae, in the collection of the British Museum; 1868.

Günther A. On some fishes from Kilima–Njaro District. Proc Zool Soc Lond 1889;(pt 1): 70–2.

Gurumayum SD, Kosygin L. *Garra tamangi*, a new species of cyprinid fish (Teleostei: Cypriniformes) from Arunachal Pradesh, northeastern India. Species 2016;17(55):84–93.

Gurumayum SD, Kosygin L, Tamang L. Ichthyofaunal diversity of Arunachal Pradesh, India: a part of Himalaya biodiversity hotspot. Int J Fish Aqua Stud 2016;4(2):337–46.

Gurung DB, Dorji S, Tshering U, Wangyal JT. An annotated checklist of fishes from Bhutan. J Threat Taxa 2013;5(14):4880–6.

Hamilton F. A journey from Madras through the countries of Mysore, Canara, and Malabar, performed under the orders of the most noble the Marquis Wellesley, Governor General of India. London: T. Cadell and W. Davies etc.; 1807.

Hamilton F. An account of the fishes found in the river Ganges and its branches. Edinburgh/London: Archibald Constable/Hurst, Robinson; 1822.

Havird JC, Page LM. A revision of *Lepidocephalichthys* (Teleostei: Cobitidae) with descriptions of two new species from Thailand, Laos, Vietnam, and Myanmar. Copeia 2010;1:137–59.

Heckel JJ. Fische aus Caschmir gesammelt und herausgegeben von Carl Freiherrn von Hügel, beschrieben von J. J. Heckel. Wien; 1838.

Heckel JJ. Ichthyologie [von Syrien]. In: von Russegger J, editor. Reisen in Europa, Asien und Africa, mit besonderer Rücksicht auf die naturwissenschaftlichen Verhältnisse der betreffenden

Länder unternommen in den Jahren 1835 bis 1841, etc. E. Schweizerbart'sche Verlagshandlung. Stuttgart. Ichthyologie [von Syrien]. Russegger, vol. 1 (pt 2); 1843. p. 990–1099.

Hollister G. Clearing and dying fishes for bone study. Zoologica 1934;12:89–101.

Holly M. Zur Nomenklatur der Siluridengattung *Macrones* C. Duméril. Zool Anz 1939;125(5/6):143.

Hora SL. On some new or rare species of fish from the eastern Himalayas. Rec Indian Mus (Calcutta) 1921a;22(33 (pt 5)):731–44.

Hora SL. Indian cyprinoid fishes belonging to the genus *Garra*, with notes on related species from other countries. Rec Indian Mus (Calcutta) 1921b;22(29 (pt 5)):633–87.

Hora SL. Notes on fishes in the Indian Museum, V. On the composite genus *Glyptosternon* McClelland. Rec Indian Mus (Calcutta) 1923;25(pt 1):1–44.

Hora SL. Notes on fishes in the Indian Museum. XXIV. Loaches of the genus *Nemachilus* from eastern Himalayas, with the description of a new species from Burma and Siam. Rec Indian Mus (Calcutta) 1935;37(pt 1):49–67.

Hora SL. The game fishes of India. Part III. J Bombay Nat Hist Soc 1937;39(pt 4):659–78.

Hora SL. Siluroid fishes of India, Burma and Ceylon. XIII. Fishes of the genera *Erethistes* Müller and Troschel, *Hara* Blyth and of two new allied genera. Rec Indian Mus (Calcutta) 1950;47 (pt 2):183–202.

Hora SL, Mukerji DD. Fishes of the Naga Hills, Assam. Rec Indian Mus (Calcutta) 1935;37 (pt 3):381–404.

Howes GJ. Anatomy and evolution of the jaws in the semiplotine carps with a review of the genus *Cyprinion* Heckel, 1843 (Teleostei: Cyprinidae). Bull Br Mus (Nat Hist) Zool 1982;42(4):299–335.

Jarocki FP. Zoologiia czyli zwiérzetopismo ogólne podlug náynowszego systematu. Warszawie (Warsaw): Drukarni Lakiewicza; 1822.

Jayaram KC. A new species of sisorid fish from the Kameng Frontier Division. Nefa J Zool Soc India 1966;15(1):85–7. [for 1963].

Jayaram KC. Contributions to the study of bagrid fishes (Siluroidea: Bagridae). 10. Systematic position of *Pimelodus chandramara* Hamilton with description of a new genus. Int Rev Gesamten Hydrobiol 1972;57(5):815–20. [for 1971].

Jayaram KC. The fresh water fishes of the Indian region. New Delhi: Narendra Publishing House; 1999.

Jayaram KC. Catfishes of India. Delhi: Narendera Publishing Houde; 2006.

Jayaram KC, Mazumdar N. On a collection of fish from Kameng frontier division. NEFA J Bombay Nat Hist Soc 1964;61(2):264–80.

Jordan DS. New genera of fishes. Proc Acad Nat Sci Phila 1919;70:341–4. [for 1918].

Katwate U, Jadhav S, Kumkar P, Raghavan R, Dahanukar N. *Pethia sanjaymoluri*, a new species of barb (Teleostei: Cyprinidae) from the northern Western Ghats, India. J Fish Biol 2016;88(5):2027–50.

Kaup JJ. Uebersicht der Lophobranchier. Arch Naturgesch 1853;19(1):226–34.

Khynriam D, Sen N. On a new species *Badis triocellus* (Pisces: Perciformes: Badidae) from North East India. Rec Zool Soc India 2013;111(4):65–72. [for 2011].

Khynriam D, Sen N. Taxonomic study of nemacheiline loaches of north east India. Rec Zool Soc India 2014;1–37. Occasional paper no. 358.

Knight JDM. *Pethia aurea* (Teleostei: Cyprinidae), a new species of barb from West Bengal, India, with redescription of *P. gelius* and *P canius* Zootaxa. 2013;3700(1):173–84.

Knight JDM. *Oreichthys andrewi* (Teleostei: Cyprinidae) a new species from Assam, northeastern India. J Threat Taxa 2014;6(1):5357–61.

Knight JDM. *Channa pardalis*, a new species of snakehead (Teleostei: Channidae) from Meghalaya, northeastern India. J Threat Taxa 2016;8(3):8583–9.

Koerber S, Litz TO, Mirande JM. CLOFFAR – Update 3 – Supplement to checklist of the freshwater fishes of Argentina. Ichthyol Contrib PecesCriollos 2017;47:1–9.

Kosygin L. *Aborichthys waikhomi*, a new species of fish (Teleostei: Nemacheilidae) from Arunachal Pradesh, India. Rec Zool Soc India 2012;112(1):49–55.

Kottelat M. Indochinese nemacheilines. A revision of nemacheiline loaches (Pisces: Cypriniformes) of Thailand, Burma, Laos, Cambodia and southern Viet Nam. München: Verlag Dr. Friedrich Pfeil; 1990.

Kottelat M. On the valid generic names for the Indian fishes usually referred to *Salmostoma* and *Somileptes* (Teleostei: Cyprinidae and Cobitidae). J South Asian Nat Hist 1998;3(1):117–9.

Kottelat M. Nomenclature of the genera *Barbodes, Cyclocheilichthys, Rasbora* and *Chonerhinos* (Teleostei: Cyprinidae and Tetraodontidae), with comments on the definition of the first reviser. Raffles Bull Zool 1999;47(2):591–600.

Kottelat M. Fishes of Laos. Colombo: WHT Publications; 2001.

Kottelat M. *Parambassis pulcinella,* a new species of glassperch (Teleostei: Ambassidae) from the Ataran River basin (Myanmar), with comments on the family–group names Ambassidae, Chandidae and Bogodidae. Ichthyol Explor Freshw 2003;14(1):9–18.

Kottelat M. *Acanthocobitis pictilis*, a new species of loach from Myanmar and Thailand (Teleostei: Nemacheilidae). Zootaxa 2012;3327:45–52.

Kottelat M. The fishes of the inland waters of Southeast Asia: a catalogue and core bibliography of the fishes known to occur in freshwaters, mangroves and estuaries. Raffles Bull Zool 2013;27:1–663.

Kottelat M. The fishes of the Nam Theun and Xe Bangfai drainages, Laos. Hydroécologie Appliquée (Paris) 2015;1–51.

Kottelat M. On Gonorynchus, Gonorhynchus, Gonorinchus, Gonorhinchus and Gonorrhynchus, and some other names of labeonine fishes (Teleostei: Gonorynchidae and cyprinidae). Zootaxa 2016;4178(3):443–50.

Kottelat M, Freyhof J. Handbook of European freshwater fishes. Cornol: Kottelat; 2007.

Kottelat M, Whitten T. Freshwater biodiversity in Asia with special reference to fish: World Bank Technical paper No. 343. Washington, DC: The world bank; 1996.

Kuiter RH. Seahorses and their relatives. Seaford: Aquatic Photographics; 2009.

Kullander SO. Taxonomy of chain *Danio*, an Indo–Myanmar species assemblage, with descriptions of four new species (Teleostei: Cyprinidae). Ichthyol Explor Freshw 2015;25(4):357–80.

Kullander SO, Britz R. Revision of the family Badidae (Teleostei: Perciformes), with description of a new genus and ten new species. Ichthyol Explor Freshw 2002;13(4):295–372.

Kullander SO, Fanf F, Delling B, Ahlander E. The fishes of the Kashmir Valley. In: Nyman L, editor. River Jhelum, Kashmir Valley, impacts on the aquatic environment. Goteborg: Swedmar; 1999. p. 99–163.

Lacepède BGE. Histoire naturelle des poisons. Paris: Imprimeur-Libraire; 1800 [Publication date: Hureau & Monod 1973].

Lacepède BGE. Histoire naturelle des poisons. Paris: Imprimeur-Libraire; 1803 [Publication date: Hureau & Monod 1973].

Lalhlimpuia DV, Lalronunga S, Lalramliana. *Channa aurantipectoralis*, a new species of snakehead from Mizoram, north–eastern India (Teleostei: Channidae). Zootaxa 2016;4147(3):343–50.

Lalramliana, Knight JDM, Laltlanhlua Z. *Pethia rutila* (Teleostei: Cyprinidae), a new species from Mizoram, northeast India. Zootaxa 2014;3827(3):366–74.

Lalramliana, Lalnuntluanga, Lalronunga S. *Psilorhynchus kaladanensis*, a new species (Teleostei: Psilorhynchidae) from Mizoram, northeastern India. Zootaxa 2015;3962(1):171–8.

Lalronunga S, Lalnuntluanga, Lalramliana. *Schistura maculosa*, a new species of loach (Teleostei: Nemacheilidae) from Mizoram, northeastern India. Zootaxa 2013;3718(6):583–90.

Linnaeus C. Systema naturae. In: Systema naturae per regna tria naturae, secundum classes, ordines, genera, species, cum characteribus, differentiis, synonymis, locis. Tomus I. Editio decima, reformata. Holmiae. 1758.

Linthoingambi I, Vishwanath W. *Oreoglanis majusculus*, a new glyptosternine catfish from Arunachal Pradish, India (Teleostei: Sisoridae). Zootaxa 2011;2754:60–6.

Lokeshwor Y, Vishwanath W. *Physoschistura dikrongensis*, a new loach from Arunachal Pradesh, India (Teleostei: Nemacheilidae). Zootaxa 2012;3586:249–54.

Lokeshwor Y, Vishwanath W. *Schistura porocephala*, a new nemacheilid loach from Koladyne basin, Mizoram, India (Teleostei: Nemacheilidae). Ichthyol Res 2013;60(2):159–64.

Mayanglambam D, Vishwanath W. *Barilius profundus*, a new cyprinid fish (Teleostei: Cyprinidae) from Koladyne basin, India. J Threat Taxa 2012;4(2):2363–9.

Mayanglambam D, Vishwanath W. *Parambassis serrata*, a new species of glassperch (Teleostei: Ambassidae) from the Kaladan drainage, India. Zootaxa 2015;4040(5):583–8.

McClelland J. Observations on six new species of Cyprinidae, with an outline of a new classification of the family. J Asiat Soc Bengal 1838;7:941–8.

McClelland J. Indian Cyprinidae. Asiat Res 1839;19(pt 2):217–471.

McClelland J. On the fresh–water fishes collected by William Griffith, Esq., F. L. S. Madras Medical Service, during his travels under the orders of the Supreme Government of India, from 1835 to 1842. Calcutta J Nat Hist 1842;2(8):560–89.

McClelland J. Description of four species of fishes from the rivers at the foot of the Boutan Mountains. Calcutta J Nat Hist 1845;5(18):274–82.

Menon AGK. Monograph of the cyprinid fishes of the genus *Garra*. Hamilt Meml Indian Mus 1964;14(4):173–260.

Menon AGK. Check list – Fresh water fishes of India. Rec Zool Soc India 1999 . Occasional paper no. 175.

Mirza MR, Saboohi N. A note on the freshwater fishes of the river Dasht with the description of *Tariqilabeo* new subgenus (Pisces, Cyprinidae). Pak J Zool 1990;22(4):405–6.

Misra KS. Teleostomi: Cypriniformes; Siluri. In: The fauna of India and the adjacent countries. Pisces. 2nd ed., vol. 3. Faridabad: Government of India Press; 1976.

Muller J. Ueber Nebenkiemen und Wundernetze. In: Archiv für Anatomie, Physiologie und Wissenschaftliche Medicin. Verbindung mit mehreren Gelehrten Herausgebern, vol. 7; 1840. p. 101–42.

Muller J, Troschel FH. Horae Ichthyologicae. Beschreibung und Abbildung neuer Fische. Berlin. 1849.

Musikasinthorn P. *Channa panaw*, a new channid fish from the Irrawaddy and Sittang River basins, Myanmar. Ichthyol Res 1998;45(4):355–62.

Musikasinthorn P. *Channa aurantimaculata*, a new channid fish from Assam (Brahmaputra River basin), India, with designation of a neotype for *C. amphibeus* (McClelland,1845). Ichthyol Res 2000;47(1):27–37.

Nath P, Dam D, Bhutia PT, Dey SC, Das DN. A new fish species of the genus *Bhavania* (Homalopteridae: Homalopterinae) from river Noadhing drainage, Arunachal Pradesh, India. Rec Zool Surv India 2007;107(part 3):71–8.

Nath P, Dey SC. Two new fish species of the genus *Amblyceps* Blyth from Arunachal Pradesh, India. J Assam Sci Soc 1989;32(1):1–6.

Nath P, Dey SC. Fish and fisheries of north eastern India (Arunachal Pradesh). New Delhi: Narendra Publishing House; 2000.

Nebeshwar K, Vishwanath W. Three new species of *Garra* (Pisces: Cyprinidae) from north–eastern India and redescription of *G. gotyla*. Ichthyol Expl Freshw 2013;24(2):97–120.

Nebeshwar K, Vishwanath W. Two new species of *Garra* (Pisces: Cyprinidae) from the Chindwin River basin in Manipur, India, with notes on some nominal *Garra* species of the Himalayan foothills. Ichthyol Explor Freshw 2015;25(4):305–21.

Nebeshwar K, Bagra K, Das DN. A new species of the cyprinoid genus *Psilorhynchoides* Yazdani et al. (Cypriniformes: Psilorhynchidae) from Arunachal Pradish, India. Zoos' Print J 2007;22(3):2632–6.

Nebeshwar K, Vishwanath W, Das DN. *Garra arupi*, a new cyprinid fish species (Cypriniformes: Cyprinidae) from upper Brahmaputra basin in Arunachal Pradish, India. J Threat Taxa 2009;1(4):197–202.

Nebeshwar K, Bagra K, Das DN. *Garra kalpangi*, a new cyprinid fish species (Pisces: Teleostei) from upper Brahmaputra basin in Arunachal Pradesh, India. J Threat Taxa 2012;4(2):2353–62.

Nelson JS, Grande TC, Wilson MVH. Fishes of the world. 5th ed. Hoboken, NJ: John Wiley & Sons, Inc; 2016.

Ng HH. The identity of *Batasio tengana* (Hamilton, 1822), with the description of two new species of *Batasio* from north–eastern India (Teleosei: Bagridae). J Fish Biol 2006;68(A):101–18.

Ng HH, Dodson JJ. Morphological and genetic descriptions of a new species of catfish, *Hemibagrus chrysops*, from Sarawak, east Malaysia, with an assessment of phylogenetic relationships (Teleostei: Bagridae). Raffles Bull Zool 1999;47(1):45–57.

Ng HH, Ferraris Jr. CJ. A new species of anguilliform catfish (Actinopterygii: Siluriformes: Bagridae) from Bangladesh and northeastern India. Zootaxa 2016;4079(3):381–7.

Ng HH, Kottelat M. A new species of *Mystus* from Myanmar (Siluriformes: Bagridae). Copeia 2009;2:245–50.

Ng HH, Kottelat M. *Chaca serica*, a new species of frogmouth catfish (Teleostei: Siluriformes) from southern Borneo. Zootaxa 2012;3258:37–45.

Ng HH, Kottelat M. *Hyalobagrus*, a new genus of miniature bagrid catfish from Southeast Asia (Teleostei: Siluriformes). Ichthyol Expl Freshwr 1998;9:335–46.

Ng HH, Kottelat M. A review of the genus *Batasio* (Teleostei: Bagridae) in Indochina, with the description of *B. tigrinus* sp. n. from Thailand. Rev Suisse Zool 2001;108(3):495–511.

Ng HH, Pethiyagoda R. *Mystus zeylanicus*, a new species of bagrid catfish from Sri Lanka (Teleostei: Bagridae). Ichthyol Explor Freshw 2013;24(2):161–70.

Ng HH, Rachmatika I. *Glyptothoraxexodon*, a new species of rheophilic catfish from Borneo (Teleostei: Sisoridae). Raffles Bull Zool 2005;53:251–5.

Ng HH, Tamang L. *Pseudolaguvia viriosa*, a new catfish (Teleostei: Sisoridae) from north–eastern India. Zootaxa 2012;3522:81–8.

Ng HH, Wright JJ. A new torrent catfish from western Thailand (Siluriformes: Amblycipitidae). Copeia 2009;(2)369–77.

Ng HH, Ferraris Jr CJ, Neely DA. The catfish genus *Erethistoides* (Siluriformes: Sisoridae) in Myanmar, with descriptions of three new species. Zootaxa 2012;3254:55–68.

Ng HH, Lalramliana, Lalronunga S. A new diminutive sisorid catfish (Actinopterygii: Siluriformes) from northeastern India. Zootaxa 2016;4105(6):546–56.

Oken L. V. Kl. Fische. Isis (Oken) 1817;8(148):1779–82 [for 1179–1182 + [1182a]].

Pallas PS. Spicilegia Zoologica quibus novae imprimis et obscurae animalium species iconibus, descriptionibus atque commentariis illustrantur. Berolini, Gottl. August. Lange, 1769.

Peters W. Diagnosen von neuen Flussfischen aus Mossambique. In: Monatsberichte der Königlichen Preussischen Akademie der Wissenschaften zu Berlin. 1852. 275–6, 681–5.

Peters W. Uber zwei neue Gattungen von Fischen aus dem Ganges. In: Monatsberichte der Königlichen Preussischen Akademie der Wissenschaften zu Berlin. 1861. p. 712–3.

Pethiyagoda R, Meegaskumbura M, Maduwage K. A synopsis of the South Asian fishes referred to *Puntius* (Pisces: Cyprinidae). Ichthyol Explor Freshw 2012;23(1):69–95.

Phurin S, Landge AT, Darshan A, Abujam SK, Das DN. New record of Danio assamila Kullander (Teleostei: Cyprinidae) from Arunachal Pradesh, India. Int J Fish Aqu Stud 2017;5(3):425–9.

Pillai RS, Yazdani GM. Two new species and two records of *Lepidocephalichthys* Bleeker [Pisces: Cobitidae] from Assam and Meghalaya, India, with a key to the known species. J Zool Soc India 1976;26:11–7.

Playfair RL. On the fishes of Cachar. Proc Zool Soc Lond 1867;(pt 1):14–7. Pl. 3.

Rafique M. Fish diversity and distribution in Indus River and its drainage system. Pak J Zool 2000;32(4):321–32.

Rainboth WJ. *Neolissochilus*, a new genus of South Asian cyprinid fishes. Beaufortia 1985;35(3):25–35.

Regan CT. The classification of the teleostean fishes of the order Synentognathi. Ann Mag Nat Hist Ser 8 1911;7(40):327–35 [Art 36].

Richardson J. Fishes. In: Back G, editor. Narrative of the Arctic land expedition to the mouth of the Great Fish River, and along the shores of the Arctic Ocean in the years 1833, 1834, and 1835; London; 1836. i–x + 1–663, 14 pls.

Roberts TR. Revision of the south and southeast Asian sisorid catfish genus *Bagarius*, with description of a new species from the Mekong. Copeia 1983;(2):435–45.

Roberts TR. A review of the African electric catfish family Malapteruridae, with descriptions of new species. Occas Pap Ichthyol 2000;1:1–15.

Roberts TR. *Ayarnangra estuarius*, a new genus and species of sisorid catfish from the Ayeyarwaddy basin, Myanmar. Nat Hist Bull Siam Soc 2001;49(1):81–7.

Roberts TR, Ferraris Jr CJ. Review of South Asian sisorid catfish genera *Gagata* and *Nangra*, with descriptions of a new genus and five new species. Proc Calif Acad Sci 1998;50(14):315–45.

Schäfer F. *Oreichthys crenuchoides*, a new cyprinid from west Bengal, India. Ichthyol Explor Freshw 2009;20(3):201–11.

Schrank F, Von P. Fauna Boica. In: Durchgedachte Geschichte der in Baieren einheimischen und zahmen Thiere. Nürnberg. 1798.

Scopoli JA. Introductio ad historiam naturalem, sistens genera lapidum, plantarum et animalium hactenus detecta, caracteribus essentialibus donata, in tribus divisa, subinde ad leges naturae. Prague; 1777.

Sen N. Description of a new species of *Aborichthys*, Chaudhuri from North–east India (Pisces: Cypriniformes: Balitoridae). Rec Zool Soc India 2009;109(part 2):13–20.

Sen N, Biswas BK. On a new species of *Nangra* day (Pisces: Siluriformes: Sisoridae) from Assam, north east India with a note on comparative studies of other known species. Rec Zool Soc India 1994;94(2–4):441–6.

Shangningam B, Vishwanath W. *Psilorhynchus konemi*, a new species of torrent minnow from northeast India (Teleostei: Psilorhynchidae). Ichthyol Explor Fres 2016;27(4):289–96.

Shangningam B, Vishwanath W. Two new species of *Garra* from the Chindwin basin, India (Teleostei: Cyprinidae). Ichthyol Explor Freshw 2015;26(3):263–72.

Shrestha TK. Ichthyology of Nepal. A study of fishes of the Himalayan waters. Kathmandu: Himalayan Ecosphere; 2008.

Singer RA, Page LM. Revision of the zipper loaches, *Acanthocobitis* and *Paracanthocobitis* (Teleostei: Nemacheilidae), with descriptions of five new species. Copeia 2015;103(2):378–401.

Singh A, Sen N, Bănărescu PM, Nalbant TT. New noemacheiline loaches from India (Pisces, Cobitidae). In: Travaux du Muséum d'Histoire Naturelle "Grigore Antipa". vol. 23; 1982. p. 201–12. [for 1981].

Sinha B, Tamang L. *Creteuchiloglanis arunachalensis*, a new species of glyptosternine catfish (Teleostei: Sisoridae) from Arunachal Pradesh, northeastern India. Ichthyol Res 2014;62(2):189–96. [1–8].

Sinha B, Tamang L. *Oreoglanis pangenensis*, a new species of torrent catfish from Arunachal Pradesh, India (Siluriformes: Sisoridae). Ichthyol Explor Freshw 2015;25(4):331–8.

Smith HM. Contributions to the ichthyology of Siam. II–VI. J Siam Soc Nat Hist Suppl 1933;9(1):53–87.

Smith HM. *Chagunius*, a new genus of Asiatic cyprinoid fishes. Proc Biol Soc Wash 1938;51:157–8.

Srivastava CB. On a collection of fishes from Tirap Frontier Division (NEFA), India. J Zool Soc India 1966;18:122–8.

Steindachner F. Ichthyologische Mittheilungen. (IX.) [With subtitles I–VI.]. In: Verhandlungen der K–K zoologisch–botanischen Gesellschaft in Wien. vol. 16; 1866. p. 761–96.

Suckley G. Notices of certain new species of North American Salmonidae, chiefly in the collection of the N W Boundary Commission, in charge of Archibald Campbell, Esq, Commissioner of the United States, collected by Doctor C B R Kennerly, naturalist to the. Ann Lyceum Nat Hist New York 1861;7(art. 30):306–13.

Swainson W. On the natural history and classification of fishes, amphibians, & reptiles, or monocardian animals. London: A Spottiswoode; 1838.

Swainson W. On the natural history and classification of fishes, amphibians, & reptiles, or monocardian animals. London: Spottiswoode & Co.; 1839.

Sykes WH. On the fishes of the Deccan. Proc Zool Soc Lond 1839;(pt 6): 157–65. [for 27 Nov. 1838].

Talwar PK, Jhingran AG. Inland fishes of India and adjacent countries. vols. 2. New Delhi/Bombay/Calcutta: Oxford & IBH Publishing Co.; 1991.

Tamang L. *Garra magnidiscus*, a new species of cyprinid fish (Teleostei: Cypriniformes) from Arunachal Pradesh, northeastern India. Ichthyol Explor Freshw 2013;24(1):31–40.

Tamang L, Chaudhry S. *Glyptothorax dikrongensis*, a new species of catfish (Teleostei: Sisoridae) from Arunchal Pradesh, northeastern India. Ichthyol Res 2011;58(1):1–9.

Tamang L, Sinha B. Two new species of the South Asian catfish genus *Pseudolaguvia* from northeastern India (Teleostei: Sisoridae). Zootaxa 2014;3887(1):37–54.

Tamang L, Sinha B. *Physoschistura walongensis*, a new species of loach (Teleostei: Nemacheilidae) from Arunachal Pradesh, northeastern India. Zootaxa 2016;4173(3):280–8.

Tamang L, Chaudhry S, Choudhury D. *Erethistoides senkhiensis*, a new catfish (Teleostei: Erethistidae) from India. Ichthyol Explor Freshw 2008;19(2):185–91.

Tamang L, Sinha B, Gurumayum SD. *Exostoma tenuicaudata*, a new species of glyptosternine catfish (Siluriformes: Sisoridae) from the upper Brahmaputra drainage, northeastern India. Zootaxa 2015;4111(5):598–600.

Thomson AW, Page LM. Genera of the Asian catfish families Sisoridae and Erethistidae (Teleostei: Siluriformes). Zootaxa 2006;1345:1–96.

Thoni RJ, Hart R. Repatriating a lost name: notes on McClelland and Griffith's *Cobitis boutanensis* (Cypriniformes: Nemacheilidae). Zootaxa 2015;3999(2):291–4.

Valdesalici S, Voort SV. Four new species of the Indo–Burmese genus *Badis* from West Bengal, India (Actinopterygii: Perciformes: Badidae). Zootaxa 2015;3985(3):391–408.

Valenciennes A. Ichthyologie des îles Canaries, ou histoire naturelle des poissons rapportés par Webb & Berthelot. In: Webb PB, Berthelot S, editors. Histoire naturelle des îles Canaries. Paris. vol. 2 (pt 2); 1835–1850. p. 1–109. 26 pls.

Valenciennes A. Ichthyologie. In: du Petit-Thouars A, editor. Atlas de Zoologie. Voyage autour du monde sur la frégate "Vénus," pendant les années 1836–1839. Paris: Gide; 1846. 1 Table + Pls. 1–10.

Vierke J. Ein farbenfroher neuer Schlangenkopffisch aus Assam: *Channa bleheri* spec. nov. Das Aquarium: Monatsmagazin für Vivaristik, Bornheim 1991;25(259):20–4.

Vishwanath W, Darshan A. Two new catfish species of the genus *Pseudecheneis* Blyth (Teleostei: Siluriformes) from northeastern India. Zoos' Print J 2007;22(3):2627–31.

Vishwanath W, Kosygin L. Fishes of the cyprinid genus *Semiplotus* Bleeker 1859, with description of a new species from Manipur, India. J Bombay Nat Hist Soc 2000;97(pt 1):92–102.

Vishwanath W, Lakra WS, Sarkar UK. Fishes of Northeast India. Lucknow, UP: NBFGR; 2007.

Vishwanath W, Shantakumar M. Fishes of the genus *Osteobrama* Heckel of northeastern India (Teleostei: Cyprinidae). Zoos' Print J 2007;22(11):2881–4.

Walbaum JJ. Petri Artedi sueci genera piscium. In quibus systema totum ichthyologiae proponitur cum classibus, ordinibus, generum characteribus, specierum differentiis, observationibus plurimis. Redactis speciebus 242 ad genera 52. Ichthyologiae pars III. Ant. Ferdin. Rose, Grypeswaldiae [Greifswald] (Part 3); 17921–723.

Wu XW, He MJ, Chu SL. On the fishes of Sisoridae from the region of Xizang. Oceanol Limnol Sin 1981;12(1):74–9.

Yazdani GM. A new genus and species of fish from India. J Bombay Nat Hist Soc 1972;69 (pt 1):134–5.

Zhang CG, Zhao YH. Species diversity and distribution of inland fishes in China. Beijing: Science Press; 2016.

Zhang CG, Zhao YH, Xing YC, Zhou W, Tang WQ. Species diversity and distribution of inland fishes in China. Beijing: Science; 2016.

Zhou W, Li X, Thomson AW. A new genus of glyptosternine catfish (Siluriformes: Sisoridae) with descriptions of two new species from Yunnan, China. Copeia 2011;(2):226–41.

Zugmayer E. Eight new fishes from Baluchistan. Ann Mag Nat Hist Ser 8 1912;10(60):595–9.

Species Index

Note: Page numbers followed by *f* indicate figures and *t* indicate tables.

Kolibri 360 GmbH
Ferdinand-Jühlke-Straße 7
99095 Erfurt, Deutschland
produktsicherheit@kolibri360.de